高等院校土建学科双语教材（中英文对照）

◆ 风景园林专业 ◆

BASICS

水景设计

DESIGNING WITH WATER

[德] 阿克塞尔·洛雷尔 著

赵晓龙 朱 逊 译

张 波 校

中国建筑工业出版社

著作权合同登记图字：01-2009-7700号

图书在版编目（CIP）数据

水景设计 /（德）洛雷尔著；赵晓龙，朱逊译．—北京：中国建筑工业出版社，2013.12

高等院校土建学科双语教材（中英文对照）

◆风景园林专业◆

ISBN 978-7-112-16011-2

Ⅰ.①水… Ⅱ.①洛…②赵…③朱… Ⅲ.①理水（园林）-景观设计-双语教学-高等学校-教材-汉、英 Ⅳ.①TU986.4

中国版本图书馆 CIP 数据核字（2013）第 250452 号

责任编辑：孙书妍 李 鸽　　责任设计：陈 旭　　责任校对：张 颖 党 蕾

高等院校土建学科双语教材（中英文对照）

◆风景园林专业◆

水景设计

[德] 阿克塞尔·洛雷尔 著
赵晓龙 朱 逊 译
张 波 校

*

中国建筑工业出版社出版、发行（北京西郊百万庄）
各地新华书店、建筑书店经销
北京嘉泰利德公司制版
北京云浩印刷有限责任公司印刷

*

开本：880×1230毫米 1/32 印张：$4^3/_8$ 字数：170千字
2014年1月第一版 2014年1月第一次印刷
定价：16.00元
ISBN 978-7-112-16011-2
（24796）

中文部分目录

CONTENTS

序

我们长途跋涉，去尼亚加拉大瀑布，惊叹这个世界的自然奇观之一。抑或是为了去欣赏 Tivoli 附近生机勃勃的喷泉，我们对小溪小河追本溯源，翻越山岭高原，在喷泉之下伸出手臂感受清凉。有很多的方式可以来体验水的光谱。设计可以提供广泛的自然元素，以及人工或艺术水景。

近距离地审视经过设计的供水设施才能发现，它们通常是能给场所带来神来之笔，能反映文化植入，赋予场地特质，或者仅仅增添了一个娱乐的可能。

创建一个室外水景就是明确地在空间、功能、概念和技术建设上提出挑战。作为规划者，我们研究与水亲密接触的不同方式，思路和设计方案可以轻而易举地形成一个概念。但是，当它作为一个实际的规划时，空间比例的设计、材料的选择或负载能力和耐用性的计算，这些都直接挑战我们有限的知识。

以上这些足以成为我们推出该系列“高等院校土建学科双语教材”的风景园林书籍的理由。这一系列旨在对第一学期的学生直观地展现风景园林，用容易理解的方式呈现主题，突出基本要素，唤起学生了解更多知识的渴望。

作者围绕水这个主题以及它的设计可能性方面去引导读者。对水的迷恋，成了魔术、娱乐和技术挑战之间相互影响制约的一个基本要素。具体的场地、扩初设计、形式语言的确定、材料的选择，这些也有所讨论。技术的细节处理被放入“技术参数”一章中。章节都伴随着视觉示例和图表作为工具，来帮助我们发展设计方案。这本书提供了有用的技巧和注意事项，以便更好地理解主题和实际应用。这一切都会带领我们走向一个成功的设计！

编辑：Cornelia Bott

FOREWORD

We travel far to marvel at Niagara Falls, one of the natural wonders of the world, or to admire the animated fountains near Tivoli. We follow brooks and creeks to their sources and, after a mountain hike, cool our arms in wooden fountains. There are many ways to experience the spectrum of water, and design can provide a wide range of nature-oriented elements as well as artificial or artistic waterscapes.

Examining designed water installations more closely shows that they are usually the element that lends flair to a site, that reflects cultural import, gives a site its prestigious character, or perhaps just provides a playful aspect.

Creating an outdoor water design presents specific challenges regarding space, function, concept, and technical construction. As planners, we study the different ways of dealing with water more closely. Ideas and design solutions can be easily integrated into a concept. However, when it comes to the actual planning, creating the design in proportion to the space, choosing materials, or calculating load capacity and durability, quickly challenge one's knowledge.

This was reason enough to launch the series of "Basics" books on landscape architecture with this subject. The series aims to present the topic to first-semester students of landscape architecture in a straightforward, easy-to-understand manner, to highlight the essential elements, and awaken the desire to know more.

The author guides readers through the entire range of the subject of water and its design possibilities. The fascination with water is presented as a basic element that oscillates between magic, recreation, and technical challenge. The specific approach to site, developing the design, finding the formal language, and correct materials are also discussed. Technical details are addressed in the chapter "Technical Parameters." The chapters are accompanied by visual examples and diagrams that serve as tools to help one develop a design solution. The book provides useful tips and notes for a better understanding of the topic, and for practical application. This all leads to a successful design!

Cornelia Bott, Editor

INTRODUCTION: THE DIVERSITY OF WATER

Landscape architecture is a rich and complex discipline. As well as being capable of creating architectural forms and structures from natural elements, it also employs the almost infinite range and diversity of nature, which is associated with immanent mystical power and holds a deep-rooted fascination.

Water has a unique position among the natural elements. The relationship that humans have with water is complex, ambivalent, and ever oscillating between too much and too little. Water is the foundation of life. Its energy, healing qualities, light, and meditative inspiration is captivating for us all, yet water also contains an element of danger. It can instill fear and awe, and drought or floods can also kill.

Working with water as a design element has involved and always will involve this field of conflict, while still flirting with the evocations, memories, or technical possibilities that water brings.

Untamed nature

Water represents untamed nature and therefore absolute purity, freedom, and infinite power; it symbolizes the opposite of a world that is fettered by technology. These issues can be expressed dynamically in the form of roaring waterfalls, powerful animated fountains, or dense mist sculptures.

Magic

Water is both dead matter and the symbol of life. It is a fundamental part of mythology and the philosophy of nature. In many parts of the world, it plays a vital role, especially where human survival depends on solving problems of water. Images of water as a magic or animated element appear in legends, songs, or symbols and can be evoked conceptually in water designs, or by adding sculptural ornaments.

Purifying relaxation

Water is also related to cleanliness because of the role it plays in washing and bathing. This can be seen in many elements based on religion, such as baptismal fonts or the fountains located outside mosques. In small footbaths, natural swimming ponds, or ornate thermal baths, water is synonymous with relaxation, play, and sport.

Image and representations

Water is also the key to wealth and power, to the extent that it can even develop into a symbol of power above and beyond the design context – as demonstrated by the ornate fountains at the foot of Roman aqueducts, the great water axis in Versailles, or the imposing river dam project in

China today. Water can be prestigious or symbolic, depending on how it is applied as a design element. Market fountains define the center of a city; shopping centers lure customers with playful water features, and waterfalls cascading down the facades of office complexes signal the importance of the institutions within.

Technical challenge

Developing solutions to technical challenges over the centuries, such as the basic water supply needed to build transportation routes or to prevent disasters, has led to a growing, substantiated knowledge of water management. Depending on the specific local challenge, there are natural aspects and site-specific technical resources, such as fountains, cisterns, or flood control structures, which can serve as technical models for designing with water.

Designing with water is always set before a diverse and complex backdrop. It references a broad range of forms, movements, and techniques, and plays with phenomena, myths, and images, thus allowing fantasy and creativity to flourish. However, the end result is ultimately what matters, that is, how well it functions as an architectural element and how it manifests the diversity and fascination of water.

Tab.1:
Examples of water elements

Type	Free elements	Recurring elements
Jetting water	Spring	Fountain
	Geyser	Water jet
	Waterfall	Cascade
Flowing water	River	Canal
	Brook	Ditch
	Runlet	Channel
Still water	Lake	Basin
	Pond	Sink
	Pool	Trough
	Puddle	Bird baths

THE FLOW OF WATER – MODELS AND EXAMPLES

Water is a much-loved design element, which can be developed in a variety of ways. This is demonstrated by an almost infinite number of designs and realized examples that reflect either the inspiration of natural landscape or artificial technological methods.

Water is in constant natural flow. Water elements can be typologically classified and their possible applications best clarified by defining various types of flow according to their character – jetting, flowing, still, and disappearing. › Tab. 1

JETTING WATER

Jetting water can be designed in a natural manner in the form of springs, geysers, mist fountains, or waterfalls. For a more artificial approach, walled fountains or large animated fountains can be used. › Figs 1, 2 and 3

The amount of jetting water is an essential part of any design concept. Other vital aspects include

- the amount of jetting water pressure (for example a trickling runlet or powerful geyser);
- the volume of water (a narrow pipe or rushing waterfall) and the number and direction of the sources (the single, straight line of a jet of water, or an animated fountain that covers a larger area);

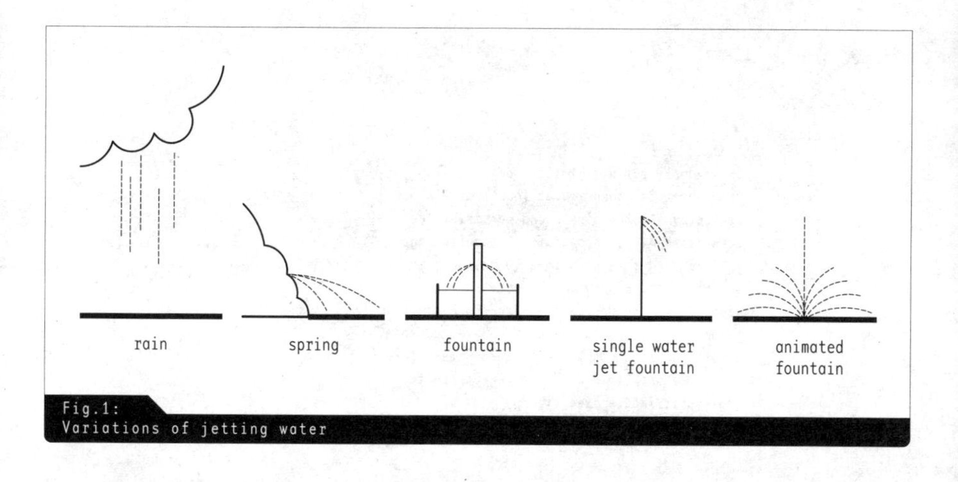

Fig.1:
Variations of jetting water

Fig.2:
Water gushes in a natural manner from between two stones

Fig.3:
Jetting water with ground-level fountains and timed movement intervals

_ the outlet's design (a small slit between stones or a beautifully forged fountain pipe), and the direct environment of the spring's source (a water source with plants or an artistically designed basin);
_ the planned intervals (a constant flow or a timed, rhythmic accentuated appearance).

\\Example:
One example for a minimalist design using jetting water is ground-level fountains without aboveground water basins and no visible components when not in operation. Stauffenegger + Stutz used this principle to design a water sculpture for a previously empty square in front of the Bundeshaus in Bern. Each of its fountains represents one of the Swiss cantons, and the sculpture's lithe, upward stream-like movements and timed dance sequences create impressive images (see Fig. 3).

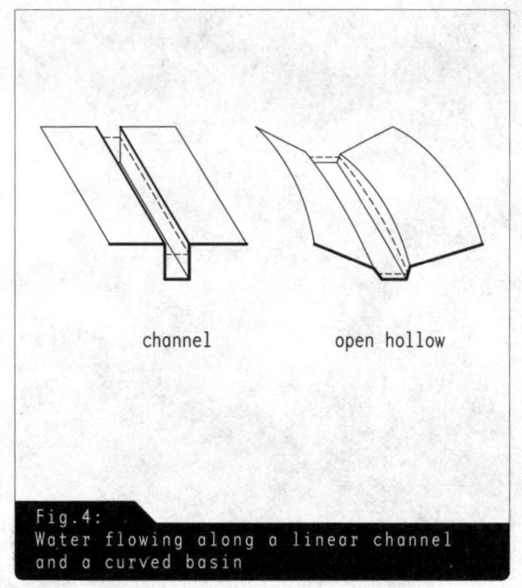

Fig.4:
Water flowing along a linear channel and a curved basin

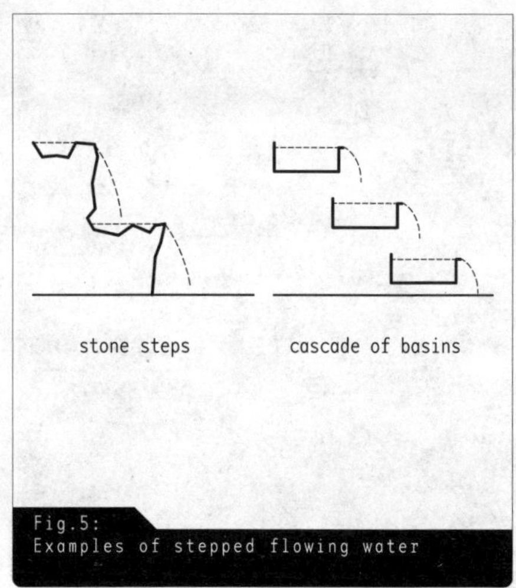

Fig.5:
Examples of stepped flowing water

Jetting water brings a fleeting image associated poetically with the naïve or childish, but also with virile buoyancy, lithe movements, lively sound, and tangible freshness.

Elements containing jetting water emphasize a point in space (for example a market fountain or a water trough near the entrance of a house), and usually create a unique space-filling character.

FLOWING WATER

Flowing water moves along longitudinal containers or in a sequence of cascading water basins. These may be landscape-inspired, such as gentle grass-lined basins, winding brooks, or stone stairways flowing with water. Artificial elements include channels, canals, or water cascades. **› Figs 4 and 5**

Possible design elements include

_ the volume of moving water, meaning the width or height of waterways and their associated speed of flow;
_ the direction of flow and how this is interrupted (for example, by straight channels, additional basins or dams);
_ the construction of the banks that contain and direct the water. **› Fig. 6**

Fig. 6:
"Natural" flowing water in a grassy basin with small gravel stones

Fig. 7:
Narrow water canals flow through stone slabs

Narrow banks covered with grass are visually subtler and allow the eye to focus more on the water's movement. Wide walls or rough gravel borders (such as loose rock fill or gabions, that is stones contained in wire mesh cages) make a clearer visual impression, accentuate the force of water, and can create a rustic aspect that grows with the size of the stones used in the design.

Flowing water in brooks or cascades convey an image of complete vigor. The movement, its interruptions, and changing speed of flow is always surprising and, for this reason, attractive to the eye. Rushing, gurgling, and splashing creates a pleasantly bright, yet gentle sound.

Most importantly however, flowing water allows linear elements (such as water channels) to develop; it connects two points in space (in the case of canals or brooks › Figs 7 and 8 and emphasizes topography (for example with waterfalls and cascades).

STILL-STANDING WATER

Still-standing water requires a hollow vessel or a drain-free, horizontal basin. These can be developed as an open landscape concept, for example in the form of shallow ponds, basins, or lakes, or in more geometric, architectural forms such as bowls, basins or sinks. › Fig. 9

Fig. 8:
A water canal as a sequence of stepped water basins

Various design possibilities could include

_ how the borders are developed (for example ground-level transitions or steps to sit on);
_ together with vegetation (completely bare, with a bordering cane break, or lush water lilies);
_ the use of lighting and its reflective qualities.

Still water reflects the sky, catches the light and shimmers. It can also absorb light, giving the impression of depth. › **Chapter Design approaches, Sensory experience**

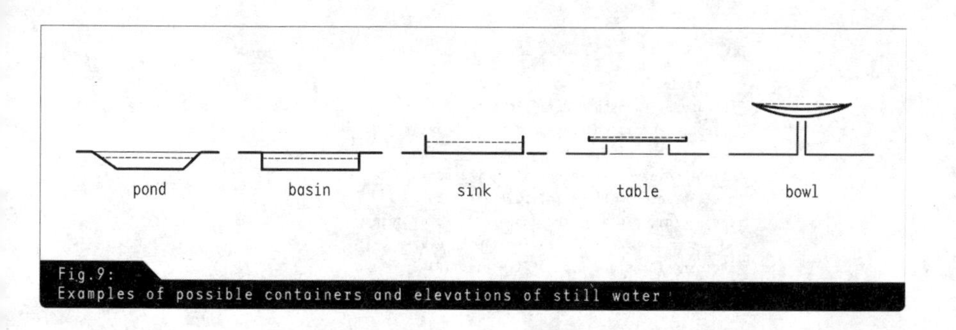

Fig. 9:
Examples of possible containers and elevations of still water

Still water expresses an inner quietude and powerful calm. It has the ability to captivate the viewer, convey something of its stillness, and inspire meditation.

Movement on the water's surface caused by wind or other external factors produces a subtle pulsing sound and a rhythmic beating of waves when they hit the perimeter or other architectural fixtures.

Still water can also architecturally mark the center of a room (for example, a reflecting basin of water in the middle of the ground floor of a building); it can clarify borders (castle moats), or serve as a subtle orientation system (a lake as a point of orientation or turning point along a visitors' footpath in a park). › Chapter Design approaches, Function Concepts that employ still water usually have greater space requirements than those mentioned above.

DISAPPEARING WATER

Disappearing water is the last section of the circulation of water. Water can disappear through drains, for example, or evaporate from impenetrable surfaces with the wind and sun, sink into porous ground, or evaporate into mist.

Design options include

_ the different possibilities of flow (decorative grates or water chutes that magnify sound); › Chapter Technical parameters, Water inflow and outflow
_ the interplay between time and speed (fleeting reflecting ponds on asphalt or water slowly dripping through a sand filter);
_ in the fortuitous and ephemeral (mist fountains or evaporation). › Chapter Technical parameters, Movement

\\Example:
In Basedow Landscape Park, Germany, Peter Joseph Lenné is working on a small body of water in the middle of a visual axis with slight pavement widenings along the banks. The green areas and trees intermittently conceal the overall picture; the expanse of water seems larger and surprises the viewer with ever-changing forms and depths (see Fig. 10).

Fig.10:
A landscape park with a long body of water with many small inlets

Fig.11:
A geometric pool of water surrounded by hedges and forest

Concepts that include evaporating water can possess a unique allure. They envelop exotic plants in wafts of mist, form ephemeral labyrinths, encase exhibition spaces in temporary, impenetrable whiteness, or create traversable clouds that float above lakes or ponds. The works created in this manner are unique and fascinating because they recontextualize familiar phenomena and give them new dimensions.

Works incorporating evaporating water are less common. They are generally very fragile, their form and range vary greatly, and their behavior is difficult to predetermine. They are often little more than experiments, which is why they are almost never seen in public space, but this aspect, as

\\Example:
In the park of Vaux-le-Vicomte, France, Le Notre places a free-standing water basin in the middle of a clearing surrounded by hedges. The formal reduction focuses on the light and shadow play of the trees, the changing shimmer of the water's surface, and hence the contrast between deliberate placement and surrounding nature (see Fig. 11).

Fig. 12:
A reflecting pool that intentionally has no center

well as its very melancholic moment of vanishing, possesses a particular allure that is worth discovering and designing.

ADDITIONAL ASPECTS

The element of water is not limited to these four options, but can also be developed in combinations (for example, in excavated springs that flow into a stream), or using specific aspects.

Absent water

The absence of water is one example. It can be demonstrated by an obvious "absence," for instance using structures and materials that are

\\Example:
In Steinberger Gasse in Winterthur, Switzerland, artists Thomas Schneider-Hoppe and Donald Judd are working on a series of simple, circular fountain basins. They play with flowing and disappearing water in a reduced and very precise manner, and surprise the viewer with a reflecting pool with no center (see Fig. 12).

Fig.13:
A recognizable "absence" of water in a dry canal bed

Fig.14:
Symbolic presentation of water using raked gravel

directly related to the existence of water (e.g. stone settings or a dry river bed). › Fig. 13 It can also be suggested symbolically, with natural elements (for example by raked areas of gravel › Fig. 14 or fields of billowy grass) or by abstract means (for instance glimmering solar panels or black asphalt surfaces).

Absent water can evoke clear and alluring images that are often contemplative, peaceful, and meditative in character.

Using absent water as a design element is recommended for installations that will be unused for extended periods of time (due to weather

\\Example:
There are beautiful examples of stone gardens in Japanese garden design, in which balanced stone settings ("islands and banks"), cut tree and hedge sculptures ("forests and solitaires"), and areas of wave-patterned, raked gravel ("water surfaces and surf") suggest dry, nature-inspired water landscapes with islands, inlets, and surf that have a reduced design, are idealized and presented in often very small spaces (see Fig. 14).

Fig.15:
Temporary ice skating rink on a reflecting basin in the middle of a post-industrial landscape

or limited hours of operation), or as an alternative to water if it cannot be used because of security issues or limited maintenance possibilities.

Winter aspect

The alluring effect of water can still be incorporated into the design even if the grounds remain in operation during periods of frost. White, soft, glazed frost can envelop bushes or trees; snow can partially conceal en element so that it becomes a reduced, abstract version of itself. On waterfalls or cascades, ice can develop into bizarre forms and images. A smooth, level

\\Example:
A 12 m wide, band-like reflecting basin was realized within a robust, former industrial structure in the Zeche Zollverein landscape park in Essen, Germany. During the winter months it becomes a popular ice-skating rink with the help of adjacent cooling units (see Fig. 15).

sheet of ice such as those found on reflecting basins or ponds can be used for winter sports such as ice-skating or ice hockey.

Even if these are merely supplementary aspects and possibilities of use, they should still be considered in relation to the local climate, if it includes long winters. The powerful image of frozen water should also be considered from a technical perspective. › Chapter Technical parameters, Winter protection

Fig.16: Tabletop fountain

DESIGN APPROACHES

Designing with water is a very individual practice influenced by a variety of factors. These include a specific handling of the element of water, the actual site, the role it plays as an architectural element in space, the functions the element needs to fulfill, its various sensory aspects, as well as the symbolic power that water can express.

DESIGNING WATER

Creating a design using water deals with its unique dynamism, the level of reference (or visual relationship) between the designed element and the viewer, the time-based experience, and the formal treatment of the containing borders (such as a bank or the edge of a basin).

Dynamics

The immanent dynamism – the manner and direction of movement – are the defining features of the design. According to the water's specific typology and the particular site, the designer needs to consider whether to use still, flowing, falling, or jetting water, and to decide on the appropriate liveliness, water amount, spatial distribution, or speed of flow.

Level of reference

The level of reference – the site of an element in relation to eye level – determines how it will be experienced and therefore defines the entire

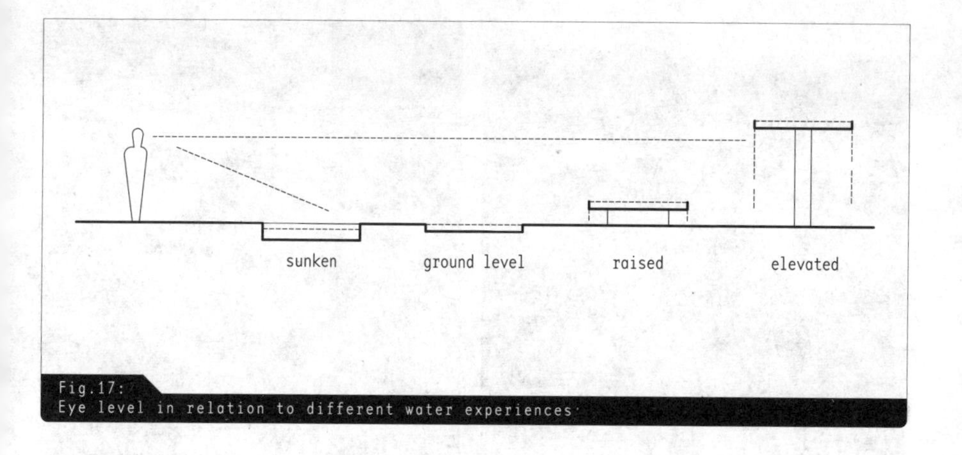

Fig.17:
Eye level in relation to different water experiences

effect of the chosen approach. A low angle of perception provides a good overview. A slightly raised angle, at knee level, offers a more tangible experience. High levels of reference provide a powerful effect at a distance. › Fig. 17

Low-lying expanses of water that offer a good overall view can be used for landscaped design, in situations where meandering banks would conceal much of the actual water's surface. They can also be used in concepts that work with visibly fluctuating water levels (such as reservoirs). Sunken bodies of water have pronounced perimeters, which could appear proportionally too large for smaller bodies of water, and limit the way the water is experienced. Raising the water to ground level avoids this problem and allows subtle transitions to be created. It requires a precisely designed and constructed perimeter with little or no water level fluctuation.

Raised elements, such as basins or bowls, can present water at either a comfortably accessible height of between 0.40 and 1.20 m › Fig. 16 or raised above eye level (over approx. 1.60 m). Due to the flat perspective and foreshortened distortion, a raised body of water seems smaller than one positioned lower. Raised elements reveal the sides and sometimes, depending on their height, the undersides of their support structure, and thus require additional design considerations of formal aspects, material, ornamental features, or the type and amount of water overflow.

Using jetting water, such as a single water jet fountain, can emphasize specific aspects by the added effect of distance, which provides a better visual experience.

Fig. 18: Installation with intervals of temporary "rain"

Fig. 19: Angular architectural interventions with calligraphic character

The experience of time

The levels of reference can be devised structurally in a timed sequence, for example with lakes of still water. Moving water offers the opportunity to develop and stage other levels of perception and temporary aspects. Spatial distribution, intensity, or timed water action › Fig. 18 can develop alternating, ephemeral images. › Chapter Technical parameters, Movement

Basin

Choosing between a natural or artificial basin is directly related to the focal point of the design. With water, this can involve light, movement, or depth, as with a reflecting pool or a freestanding single jet fountain. However, it can also develop out of the related context, the contrast and dialog between water and architecture.

With the exception of mist fountains, water can only be developed together with a basin element (such as a cistern, a collecting basin, or invisible narrow channels), and its shape plays a large role in the overall character of an object. › Chapter Technical parameters, Designing the perimeter Examples can be seen in elements and structures › Fig. 19 placed in the body of water, in ornamental fountain basins, or figurative decorations. › Fig. 20

Fig.20:
Figures framing a fountain

Fig.21:
Former water clarifier in the center of a new park landscape

UNIQUE FEATURES OF A SITE

A water design cannot be created on its own. It is one part of an architectural complex and exists within a complex spatial and thematic relationship.

The site plays a vital role in determining the effect of a designed element. The same fountain trough appears very differently from site to site – placed in the middle of a cobblestone square, in a courtyard surrounded by walls, in an overgrown bush garden, or in an open landscape. The site and its individual qualities, its overriding relationships, historical layers, and spatial formal context are essential to a coherent concept and the designed qualities of an element.

Unique features

There is great design potential in these supposed restraints. With enough financial resources almost anything is technically and creatively possible – a freedom that can result in arbitrary and interchangeable designs. It is important to search for distinctive ideas that make sense in regard to design, and not only to budget; to engage committedly with the site and its potential, and using this experience to develop individual images and sustainable technical solutions.

Overriding context

Urban structures can supply important points of reference. Significant pathways or neighboring architectural use can predetermine the site

Fig.22:
A large water crater with a jetting geyser

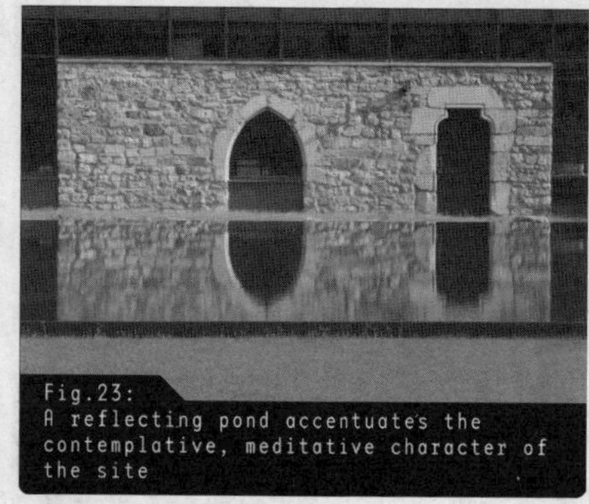

Fig.23:
A reflecting pond accentuates the contemplative, meditative character of the site

of an element. Overriding prerequisites such as rainwater management or basic design plans begin influencing the formal outcome as early as the concept stage.

Historical layers

Structures related to former use and the visible signs of a specific memory of the topography can be additional points of reference. They provide both a very unique formal character, and the additional narrative aspect of "tracking down history." The ruins of fountains or canals can be reused, overgrown streams cleared, or historical relics such as a former industrial history can be incorporated into a new context. > Fig. 21

Spatial formal concepts

The site's spatial situation, its architectural structure and existing vegetation are the criteria for the overall design in detail.

\\Example:
In Duisburg Nord landscape park, designed by Latz and Partner, the former clarifier of an abandoned steel works was cleaned and filled with water and blossoming cane break. Today it provides a distinctive focal point for the grounds (see Fig. 21). Europe's oldest artificial diving center with a newly built artificial underwater world is located in the nearby gasometer.

\\Example:
In the courtyard of an abandoned industrial complex in Berlin's Friedrichshain district, Gustav Lange placed a monolithic block of limestone, laced with fine water runlets. The block's form and proportions create a clear, individual, and powerful presence in contrast with the surrounding brick facades. The still water and slow-growing moss and ferns add poetic depth and are a charming contrast to the surroundings (see Fig. 24).

Fig. 24:
Monolithic limestone block laced with runlets placed in an architectural context

In Bad Oeynhausen, for example, powerful arrangements of water elements in space create an identifiable center point in what was once a vacant site. › Fig. 22 Their presence and beauty can echo and emphasize the language of a specific dialog, › Fig. 23 or provide their own unique response to a discourse with the surrounding architecture. › Fig. 24

Whether the formal exchange is based on a landscaped, nature-inspired discourse, or one that is more structured and architectural is influenced by context. A large, sprawling site can take a landscape-oriented response. Clear spaces, such as courtyards or other close spatial situations are more easily managed using an architecturally formal language and abstraction.

\\Note:
General information and additional design ideas can be found in *Basics Design Ideas* by Bert Bielefeld and Sebastian El khouli, Birkhäuser Verlag, Basel 2007 or in *Opening Spaces* by Hans Loidl and Stefan Bernhard, Birkhäuser Verlag, Basel 2003.

Fig.25:
Modern columns as drinking-water fountains in front of a sunken pool

Fig.26:
Rainwater pipe with an added splash-guard

FUNCTIONS

Functional requirements and specific goals can be achieved in a design involving water by using clever combinations and arrangements. In addition to the many aspects of water supply and removal, there are also those of recreation and sport to consider.

Drinking water

The original function of fountains was to supply drinking water. The development of a comprehensive public drinking water system, which ultimately included individual apartments, replaced the original function of public fountains as a town or city's water supply. Today, often filled with industrial water, they serve purely decorative purposes. Drinking fountains are more popular in warm climates, where they are connected to the drinking water supply and are common in inner cities or near recreation centers and sports facilities. › Fig. 25

Rainwater management

Precipitation that lands on sealed surfaces such as roofs or streets needs to be collected and properly disposed of in order to protect the edifice for as long as possible. Rather than directing this rainfall into the general sewage system, it is more sensible financially and ecologically to keep it on the ground where it fell and direct it back into the natural water cycle by means of evaporation or infiltration. Rainwater drainage pipes are available for this purpose, as well as collecting channels › Fig. 26 conduits, turf troughs, › Fig. 27 ponds, or underground collecting basins. There is a great amount of freedom regarding the integration and design of these technical constructions in outdoor installations. These aspects of sustainable water management are particularly significant for new developments, where a comprehensive, broad system of water management can be realized from the onset.

Fig. 27:
Rainwater retention in a grass trough by means of diagonally inserted sheets of steel

Fig. 28:
Collecting and warming basins in front of densely planted fields

Irrigation

Collected precipitation and spring water can be used to irrigate gardens and plantations. Studies have shown that water that has been exposed to a few days of sunlight and is warmed by the sun is more effective for successful plant growth. › Fig. 28 The basins, troughs, and channels needed to collect, warm, and conduct water can be integrated well into the overall design and, by exploiting synergies, can often comprise the sole element of water used.

Recreation and sport

Water is very important for recreation and sport, particularly in public parks. Given the right conditions, existing elements can be developed and integrated into an outdoor installation. One example here is the English Garden in Munich. During the planning phase, von Sckell creatively integrated the existing Eisbach into his design concept by adding waves and a landscape-like course. Today, in addition to being recognized for it formal qualities, it is a popular bathing and surfing area. › Fig. 29

Even landscaped artificial bathing areas with a nature-oriented structure and water conditioning can be easily integrated into green areas. They are however only suitable for low or average operational demands. › Fig. 30 Higher usage requires reinforced basins with an electronic water conditioning system. › Fig. 31

Fig. 29: Surfing an artificial wave in Munich's Eisbach

Fig. 30: Swimming pond with a artificial basin in the bathing section

Fig. 31: Swimming pools of varying depths of water

Fig. 32: Play area with hand water pumps, splash grounds, and toy digger

Aquatic recreation areas work with water in a variety of forms. › Fig. 73, p. 61 They make a production of the supply of water with pumps, Archimedean screws, and waterwheels; dramatize distributing and directing water using channels, cisterns, or wind vanes; and create a design in conjunction with water, sand, and mud. › Fig. 32

Enclosures and visitor orientation

Water can also serve as an impeding barrier. Wide, deep moats such as those surrounding castles can replace walls or fences. They have a similarly obstructive function yet do not interrupt the visual relationship between different areas of the grounds. Bridges and jetties situated above

Fig.33:
A moat, together with jetties and bridges directs visitors

Fig.34:
Water wall as spatial and acoustic enclosure

bodies of water link with pathways and, together with the obstructing bodies of water, can form a planned orientation system. › Fig. 33 The center and focal points of orientation systems in historical parks or modern amusement parks are often defined by large bodies of water that sometimes cannot be crossed, such as lakes or fjords. Visitors are led along prescribed routes to significant sites and attractions.

In contrast, elevated animated fountain blocks or water walls work with obstructing views. They form a protective facade that conceals undesired functions and elements. The sound they create can pleasantly block out noise from a neighboring street. › Fig. 34

SYMBOLISM

Water is rich in symbolic value and has powerful religious roots. This phenomenon has developed over the centuries and is still true today, even if at the unconscious level.

Symbol

Depending on context, the form and movement of water can represent serenity, refreshment, vitality, or wealth. It is a cross-cultural symbol of life and temporariness. Water not only literally ensures survival; it also symbolizes human intellect and spirituality. The moon, water, and femininity are closely related in terms of symbolism.

Water in religion

All three monotheistic global religions were established in dry climates – so, naturally, nature's religious significance was closely associated

Fig.35:
Oriental garden with animated fountains and canals

Fig.36:
A simple fountain near a cloister

with the element of water from their very beginnings. The purifying quality of water is often glorified: in Islam in the form of a ritual ablution, or washing before entering a mosque, or in the Hindu faith by bathing in the Ganges River. Almost every Jewish community has a Mikvah, or a ritual bath with clean, flowing water. But only those who submerge themselves fully are ritually cleansed.

For a long time, the Christian faith practiced the baptism ritual by completely immersing the body by either dipping it into water or pouring water over it. In Western societies, this practice has been replaced by dripping water onto the forehead of the person to be baptized. Baptism symbolizes a devotion to Christ and acceptance into the church. It symbolizes death and resurrection. In Catholic and Orthodox churches, holy – consecrated – water plays a significant role generally.

SENSORY EXPERIENCE

Many elements possess – beyond the purely functional level – a unique characteristic to which the design should specifically respond. Vegetation, for example, can introduce the factor of time as a fourth dimension. In harmony with the seasons, plant growth and changes in colors and shape will transform the look and mood of a landscape design, year after year. Water's unique characteristic exists in the diverse ways it can be experienced with the senses.

Designing with water is more expressive if the sensory experience is integrated into the concept and recontextualized. It is difficult to say which of the senses is most or most intensely stimulated by water. But ultimately a harmonious, well-balanced interplay is the key to a successful composition.

The sense of sight, color

Pure water is clear and transparent. It has the ability to capture light. Water drops, reflecting pools, and waves all become prisms in which light is refracted and fractured into an infinite diversity of shimmering sparkles. Water can be colored by sediments, solutions, or emulsions.

The addition of air bubbles makes water less transparent, and more white or opaque. The air dissipates as soon as the water settles and the effect disappears. This is what forms the white crests of waves or the soft, white water sculptures made by foam guns or "schaumsprudlers." › **Chapter Technical parameters, Movement**

Material or sediments picked up by flowing through soil or stones, for example, can also color water. The brown coloring often seen around the outlets of moors comes from the humic acid eluviation commonly found there. Colors derived from solutions are durable and remain stable for a longer period of time, but they are pale and cannot be controlled.

Wild, rapid flowing water tears sand and stone from the riverbed and propels it forward. These sediments color the water yet with little of the transparence we know from the fresh, green color of mountain streams. When the current subsides, the sediments settle and, with it, the color.

\\Example:
Water represents life, death, and resurrection. Fountains with contained, serene designs are placed in Christian cemeteries or nearby cloisters, for example in the form of thin jets of water, stone fountains, or small labyrinthine waterways (see Fig. 36). In oriental gardens, water represents the four rivers of paradise (see Fig. 35).

Fig.37:
"Kneipp" basin with handrail and various ground coverings

However, a body of water's "color" usually comes from its surface reflections of the surrounding environment, or from objects below the water's surface. The way we perceive these colors depends on the angle of view and refraction between the air and the water's surface, and from the difference in brightness. The deeper the water, the "deeper" the color appears – an effect that is enhanced by natural deposits that form on the river's sides and bed. The dark, almost indistinguishable floors of pond contribute to the reflecting effect on the water's surface. Light-colored floors, such as those in swimming pools, make the water look clear, bright, and transparent, and decorations on the pool's floor can therefore easily be seen. › Fig. 38

\\Example:
"Kneipp" basins are for bathing knee-deep in cold water. The up and down movements and alternating between warm and cold water are known for their rejuvenating effects, which can be enhanced by using tactile floor coverings such as gravel (see Fig. 37).

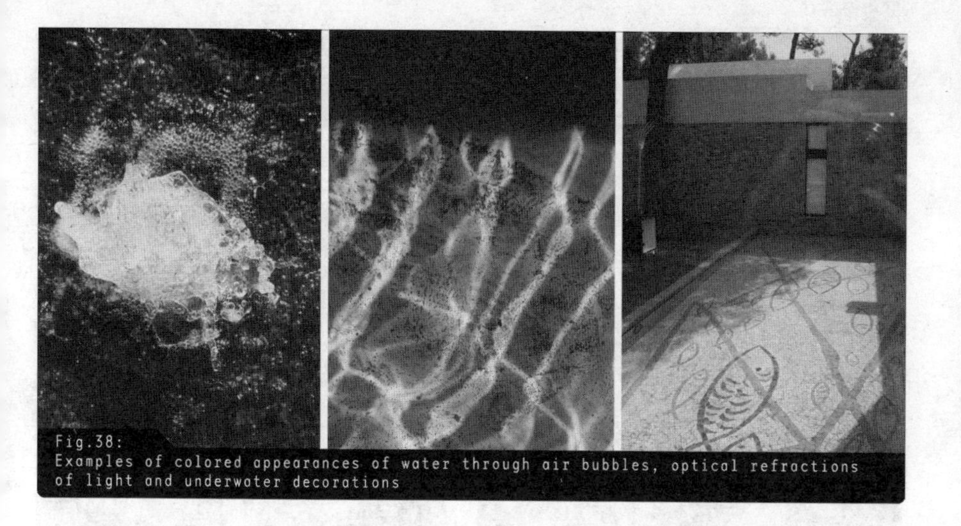

Fig.38:
Examples of colored appearances of water through air bubbles, optical refractions of light and underwater decorations

The sense of taste

Pure water does not have a flavor of its own. Eluviations and solutions from the surrounding soil or stones provide the water's flavor. Spring water can be tapped and used, for example, in drinking fountains.

The sense of smell

The "scent of water" is produced in the same way as its taste, through soluble additives or the scent of moist materials. In order to perceive the scent, aromatized water has to be released into the atmosphere as a vapor or spray. The fresh, slightly mineral scent near irregularly falling waterfalls is created in precisely this manner.

The sense of touch

Water can be directly physically tangible by diving into it, or indirectly through exposure to steam or atomization. Temperature and air humidity can relativize the effect. Atomized water, for example, is perceived as pleasantly refreshing in a dry, hot atmosphere, but in a cool, humid atmosphere it seems unpleasant, cold, and disagreeable.

The sense of hearing

The sound of water, with its highs and lows, rhythm and changing tone, reflects many musical qualities: the roar of a mountain river, the full, powerful gurgling of fountains, the muted pulsing of schaumsprudlers, or the mechanical falling of a single water drop. Pitch and sound quality depend on the amount of water, its speed and the resonant body upon which it falls. Thus, the speed of flow, the type of waterway, the shape of the

Fig.39:
A musical "Water Organ" with animated fountains, cascades, and calm water basins

Fig.40:
A distinctive setting for a water installation

fountainhead, the dimensions, height, and depth of the surface of impact, and the frequency of the pump cycle all influence the particular sound of a water installation.

\\Note:
Due to the number of influential factors, sound is as difficult to control as the fall of water, and should be tested and adjusted during the construction phase. This can be done by changing the impact flange (e.g. using different angles through which water enters), the resonant body (e.g. stone instead of wood), or speed of flow.

\\Example:
The many animated fountains, cascades, and water basins of the "Water Organ" in the gardens at the Villa d'Este gardens in Tivoli, Italy, are visual reminders of an actual musical instrument, while changing water pressure, the various forces of the jet streams, and the differentiated surfaces of impact serve as acoustic reminders (see Fig. 39).

TECHNICAL PARAMETERS

After finding a concept for a water installation based on the existing site and one's own individual inspiration, the designer must expand and develop the idea into a coherent and sustainable design that is conscious of the available technical and financial possibilities.

The questions and possible solutions are as diverse and individual as the chosen approach, and require the imagination and resourcefulness of the planner down to the last detail. Issues of the water's availability and source, the type of basin, the water's direction, movement, and overall scenographic design are paramount in this regard.

WATER AS SCENOGRAPHY

Source

The first thing to be established before considering a water installation is whether sufficient water in the required quality is available during the hours of operation. Water can be tapped from springs, lakes, rivers, rainfall, groundwater, or the local mains. It can then be collected in an intermediate reservoir and supplied to the planned design when needed.

Natural springs of sufficient capacity with natural outflow points are an ideal and affordable basis for a water design. Water quality is determined by the soil horizon through which it flows. Building a covering over the mouth of the spring protects it from impurities. This method can also combine several springs and, given sufficient reservoir volume, help compensate for the varying capacities of the different springs, as they often fluctuate over the year.

Water can also be obtained from aboveground sources such as streams or lakes with pipes, pumps, or water engines. Water mills, with their typical dams and tailraces, are a familiar example of this method. The procedure ensures a consistent supply, but the exiting water often requires extensive treatment.

Precipitation such as rain or snow can be collected year-round from roofs, town squares, or other sealed surfaces. The amount collected and how and when it is distributed is determined by the local climate, and can vary greatly. This can mean water supply shortages for the installation, especially in summer, or during a prolonged dry spell. The collected downfall is stored in aboveground basins or underground cisterns of the appropriate dimensions before being tapped and seasonally distributed to other uses. Rainwater is usually of high quality but can be contaminated by impurities from runoffs from neighboring sealed surfaces.

Groundwater can be found in the hollow spaces of pervious rocks and in sub-surface soil layers. It can be easily pumped from porous grounds such as gravel or sand, thus guaranteeing a consistent supply. Exposing an aquifer by large-scale excavation, as in gravel mining, creates permanent lakes of high-quality water.

A fairly simple way of obtaining water is to use the existing drinking water system. The water supply is usually consistent and of high quality and needs no additional treatment or conditioning. However, water bills can be costly in the medium term.

Depending on the design in question, it may also be possible to combine different water sources. The decisive factors include local availability of water, existing reservoir possibilities, ensuing cost, and desired quality.

Structure

A water installation › Fig. 41 is supplied with water via an inflow and outflow, occasionally in conjunction with a buffering reservoir. › Chapter Technical parameters, Water inflow and outflow Water can be presented alone (e.g. as a free-standing fountain) or together with a fabricated basin (e.g. trough fountain). A basin is a concave form that is either visible (basins, bowls, or hollows) or not visible (under a grating or in sub-surface pump chambers), depending on the design. › Chapter Technical parameters, Liners and basins Additional interior hydrologic cycles can be planned for moving water (such as purifying circulations or animated fountains). › Chapter Technical parameters, Movement

Water quantity

The quantity of water required is determined by the volume to be filled (fountain basins or ponds), the concealed installations (connecting

\\Note:
In most countries, any interventions or changes made to existing water is subject to strict laws. This also applies to the delivery or introduction of groundwater. The time needed to process requests and issue licenses is often lengthy, which needs to be taken into account during the project development.

\\Tip:
It is best to use drinking water in designs that include direct physical contact with water. Designated drinking fountains and installations, such as water play areas that invite direct contact with water, have to use drinking water for reasons of hygiene.

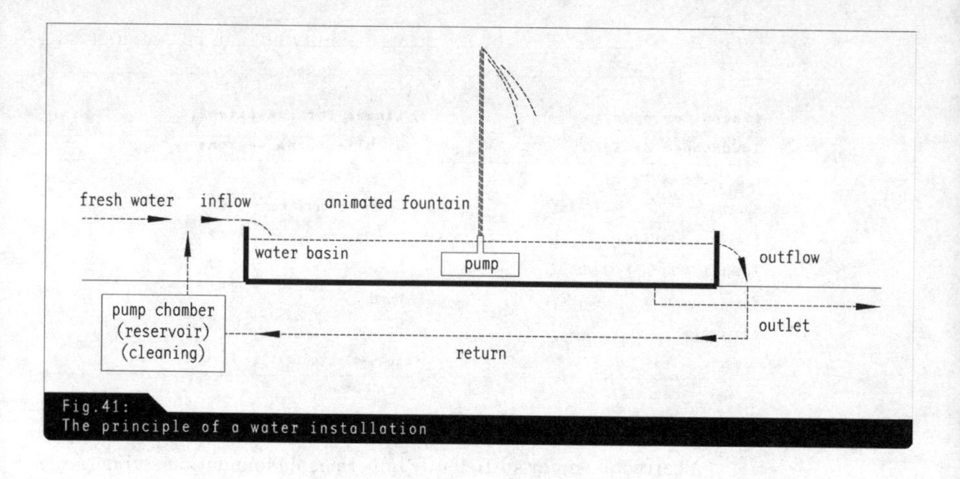

Fig.41:
The principle of a water installation

pipes or buffering reservoirs), as well as temporary quantities needed for cascades or animated fountains.

A water inflow is temporarily required to completely fill the installation, and permanently required to compensate for water loss. Loss results from regular use, for example, from spilled water, animated fountains, replacing some water in order to maintain quality, or by evaporation.

LINERS AND BASINS

Water does not tolerate errors in concept or construction. It is incredibly precise when it comes to finding the tiniest permeation, passing through it until it reaches the next impermeable area, where it stops.

\\Note:
Loss by evaporation is caused by intense exposure to sun and wind. This can result in a loss of one centimeter per day in open bodies of water, even in temperate climates. To counter the effect, make sure there is some shade and protection from wind on the site you have chosen.

Tab.2:
Different liners

Liners for open, landscaped designs	Liners for consistent, architectural designs
Clay/silt	In situ concrete
Bentonite	Concrete component
Flexible membranes	Plastic
Tar-bitumen roof sheeting	Steel
Mastic asphalt	Wood
Shotcrete	Masonry
Plastic	Natural stone

A correct basin or sufficiently impermeable foundation soil is rarely available. Thus, the decisive factors in a successful and durable water installation include choosing a durable liner; a precise, detailed development of the perimeter and transitions; and a sound connection between individual components.

Depending on material and construction, each type of liner is unique in character beyond its technical advantages and disadvantages, and can support the formal requirements of a concept to a greater or lesser degree. › Tab. 2

The desired look, the design context, form and dimensions, the energy of the water's movement, the way in which the body of water will be used, and finally, the soil foundation determine the type of liner required. The top edge of the liner must always be continuous and sit firmly above the desired maximum water level. › Fig. 42 and Chapter Technical parameters, Designing the perimeter

The soil foundation needs to be sufficiently compacted and able to support the weight of the future body of water. Any later resettling can result in damages to the liner and perimeter design. As a rule the following methods are used:

Clay/silt

Clay is the oldest liner method. 30 cm of clay or silt with low water permeability ($k \leq 10^{-9}$) is applied to a stone-free, stable, and profiled foundation; it is then compacted and covered with a protective layer of gravel sand. A slope of up to 1/3 can be developed using this method.

The liner layer is delivered as dry bulk, unloaded, and applied evenly. Adobe bricks and prefabricated clay elements are also good alternatives for smaller areas. They are applied in several layers and then compacted.

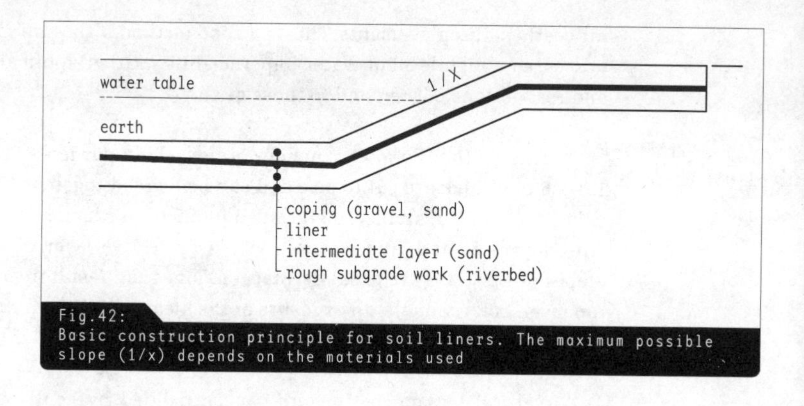

Fig.42:
Basic construction principle for soil liners. The maximum possible slope (1/x) depends on the materials used

During construction, the clay must maintain the correct degree of moisture to ensure a secure seal.

The natural character of clay and silt guarantees a high level of emotional acceptance on the part of future users. The material can also be easily disposed of if the installation has to be demolished. However, if the water supply is turned off or stops, and the material is allowed to completely dry out, deep cracks can appear. These may cause permanent damage, as do heavy use or profuse root growth. Due to their swelling property, liners made of clay or silt have a self-repairing power in smaller installations. This quality, together with adding gravel and sand to fill existing small cracks, allows the material to automatically compensate for slight damage to the liner or to absorb marginal shifts in the subsoil, without additional corrective measures. On the other hand, it is more difficult to integrate or make penetrations in clay liners; these usually require additional, wide rims of flexible membrane.

This liner method suits near-natural bodies of water that are permanently filled with water, are not heavily used, and have few installed fixtures, such as garden ponds.

Bentonite

One special form of clay liner is bentonite, a highly absorbent stone made of clay minerals, which is milled into the soil level as a mealy powder before being compacted and covered with a protective layer of gravel or sand. Some manufacturers offer alternatives to the loose admixture in the form of prefabricated mats, which can be laid onto the soil.

Bentonite increases the soil's actual impermeability to water —making the ground an additional liner layer and thereby eliminating the need for costly soil replacements. This is a good method if the site's soil already possesses a high level of water impermeability. Other application options and restrictions correspond to those of clay.

Flexible membranes

This method is most common in household gardens. Plastic sheets 1.5–2.5 mm thick are cut to size and can be bonded together. They are laid on a pre-modeled, stable, fine-grained foundation ground, on top of a leveling layer of sand, and covered with a protective layer of gravel sand. Slopes of up to 1/3 are possible. Steeper slopes can result in erosion of the top layer, and gradually to exposure of the membrane due to its sensitivity to pressure and ultraviolet rays.

Natural-looking perimeters can be molded by gently warping the membrane, as can also be done with clay liners. Connections with fixtures or penetrations can be made watertight by using flanges or terminal strips. Subsequent repairs are only possible to a certain degree.

The advantages of flexible membranes include availability, fast and easy installation, the ability to seal even very permeable foundations, and the relatively economic price.

This method is rarely used in larger technical constructions such as rainwater retention basins or designed installations for smaller landscaped pond installations.

Tar-bitumen roof sheeting

Tar-bitumen roof sheeting as a liner is similar to the flexible membrane. The tar-bitumen roof sheeting consists of sheet material coated with bitumen on both sides. The materials needed are easily available and

\\Note:
Check the ultraviolet and root resistance when choosing the membrane. The membrane should be lying flat and smooth without debris before applying the covering material, because this is where tears can develop. If there are plants at the site such as cane break or bamboo that have aggressive root growth and can easily penetrate the membrane, then a stronger cover, a higher-quality membrane, or additional root resistant plastic liner should be used for the vegetation area.

\\Tip:
Piercing the stone-filled sheet asphalt for pipelines or outflows is best done by casting a flange directly into the asphalt. To connect the construction to other architectural structures, countersink a seam and seal it with asphalt. Smaller architectural elements such as stairs can directly mounted on the liner.

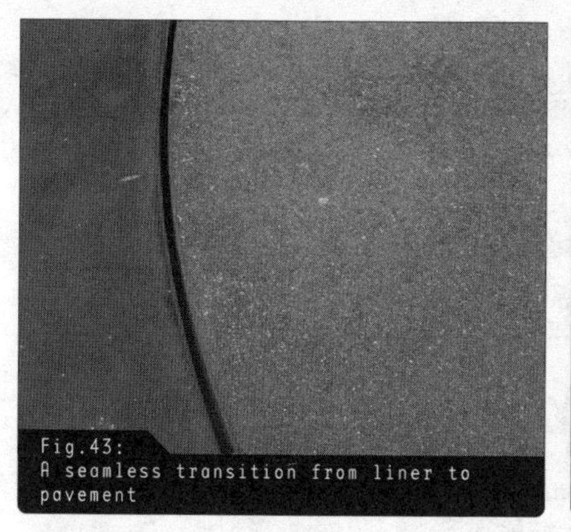
Fig.43:
A seamless transition from liner to pavement

Fig.44:
Mixed construction method with asphalt, concrete and steel

simple to use. However, they are not ultraviolet or root resistant, which means that this otherwise economic liner method can only be used for temporary installations.

Mastic asphalt

Mastic asphalt is a dense bituminous mass that is fluid when hot, and can be poured. Two 1 cm thick coats are applied to a graduated foundation ground with an anti-freeze layer, mineral substratum, and a layer of bitumen binder. This method allows for a slope of 1/2.5 and steeper. It is a durable and stable liner method and can be added to later and resealed. The procedure can only be done using the correct technique. It is a complex process and relatively expensive for small installations.

Stone-filled sheet asphalt is a suitable alternative to this method. It is more economical to produce and to install, but does not yield the same dense porosity as mastic asphalt and therefore calls for a more technically elaborate installation along the perimeters.

The stone-filled sheet asphalt liner method is recommendable for large, uneven areas, for a porous but stable foundation, and for an installation that is heavily used such as an inner-city pond installation. It allows for a smooth transition to the pavement of the surrounding pathways and can therefore be used to build shallow, traversable reflecting pools. › Fig. 43

In situ concrete

The cast-in-place method requires the water installation to be built on a stable foundation, using high-quality, watertight cement, preferably reinforced. With the correct formwork, all slope angles and exact forms can be cast on site. › Fig. 40, p. 36 Producing these components directly where they will be needed calls for a greater amount of tolerance in the original design due to the building site's unique work conditions, the specific construction of the formwork, and the shrinkage of the cement. Visible areas can be fabricated as smooth, exposed concrete, or processed as masonry, painted, or tiled.

Concrete is nontoxic and therefore safe for plants and animals. In the early stages, water runoff can alter the water's pH, but changing the water can rectify this problem.

This method is used mostly for larger architectural, geometric projects such as swimming pools, water canals or sinks, and if the installation will be used heavily. › Fig. 45

Concrete components

Concrete components are manufactured in a factory, delivered to the construction site, and installed on a prepared foundation ground. Large installations can consist of several components or, especially when building large base plates, can be added to on site using the cast-in-place method.

With assembled constructions, the joints must be worked and sealed with extreme care. Besides ensuring a high quality of concrete, industrial fabrication also guarantees very precise building elements, which affects dimensional tolerance as well as the way the surface can be developed. › Fig. 46

\\ Note:
Concrete segments larger than 5 m require expansion joints, which can be made watertight by inserting liner tape. Since these joints influence the future appearance of the water installation, it is important to consider the design as well as the technical aspects of these sections, for example, whether or not to place the joints at regular intervals, and so on. Even after treating the surface by stabbing or elution, the reinforcements still need to be protected. For concrete hydraulic engineering constructions, a large concrete cover is recommended to protect steel reinforcements from corrosion, meaning that many constructions will need to be larger than structurally required.

Fig.45:
Trough fountain made of in situ concrete

Fig.46:
A precision-built concrete component

The available transportation method and capacity is the only aspect that can restrict the components' dimensions and forms.

The prefabricated elements are called cast stones, and their visible surface is stone cut or designed by stabbing, granulating, sandblasting, etching, water blasting, or sanding. Concrete components are popularly used as monolithic elements or as essential components for architectural designs when high precision or a specific surface structure is required, or when large quantities have to be produced or manufactured on site, away from the construction area. This method is comparable to natural stone, but is generally more economical.

Shotcrete

Shotcrete is a specific form of concrete engineering. Special concrete in thick fluid form is delivered to the construction site in a closed line and sprayed with a pneumatic gun onto a prepared surface consisting of cast forms, soil, or other components. The impact pressure of this seals the ground soil. Reinforcements might be required, depending on how the water installation will be used and the quality of the foundation ground.

This method is recommended for installations planned for a changing topography, for installations with several links to architectural structures, or for installations that will need to withstand heavy use. It is similar to the mastic asphalt method, and can be transported by pipeline if the construction site is difficult to access.

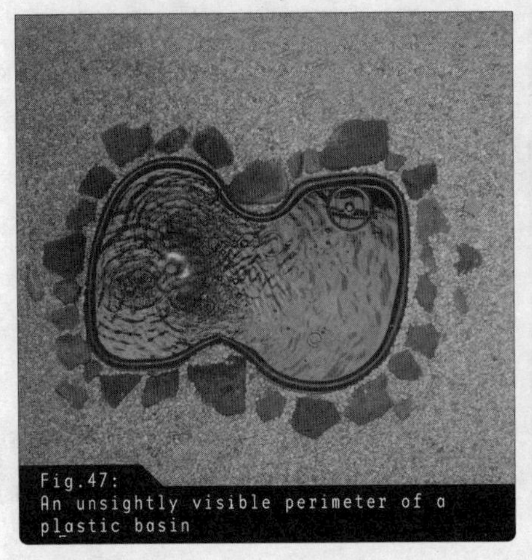
Fig.47:
An unsightly visible perimeter of a plastic basin

Fig.48:
Plants in a steel water trough with very thin walls

Plastic components

With this method, plastic- or fiberglass-reinforced synthetic resin is industrially manufactured in the required form and lowered on site into an excavation pit according to the manufacturer's instructions. Large basins often call for additional stabilizing backfill, for instance, of lean concrete. Plastic components are readily available for small- to mid-size installations, some of which can be reused. However, is difficult to conceal the fiberglass basin well, making it complicated to aesthetically integrate the structure with the surrounding grounds. › Fig. 47

This method is common for temporary installations and smaller decorative pools, for example, as an underground construction that is concealed by stones or fully visible as aboveground swimming pools.

Steel

Steel constructions are used for fountain installations, decorative pools, Kneipp footbaths, swimming pools, and installations that will be heavily used for long periods of time, such as dams, water chutes, or play areas. The optical litheness of steel is always surprising. › Fig. 48 It is highly durable and can be worked and installed very accurately, which is useful when building a reflecting pool according to specific, precise requirements.

As a rule, non-corrosive stainless steel is used, but this material can look sterile. This problem can be rectified by a variety of surface treatments including sandblasting, painting, or powder coating the surface. Paint will wear and peel after a period of time and become unsightly.

Fig.49:
Thick walls of steel for a fountain with fine jets of water

Fig.50:
Historical fountain installation built completely of wood

Galvanized and crude steel are increasingly being used to build fountains and perimeters, as well as weatherproof construction steel, a stainless-steel alloy with an interesting corrosive, weathered-looking “rusted” surface. › Fig. 49

Wood

Wood has a long tradition as water basins, particularly in forested areas. Pipelines and troughs can be fashioned from entire tree trunks. Long, box channels, densely placed together, is a method used to build overshot watermills, the wide characteristic reflecting pools for wash yards, and impressive fountain installations. › Fig. 50

These variations can be used for architectural forms and can therefore be compared with concrete construction methods. Permanent structures

\\Tip:
Even weatherproof construction steel will eventually corrode if it is kept wet or is occasionally wet over a long period. It is recommended to design the walls thicker than structurally necessary in order to avoid medium-term loss.

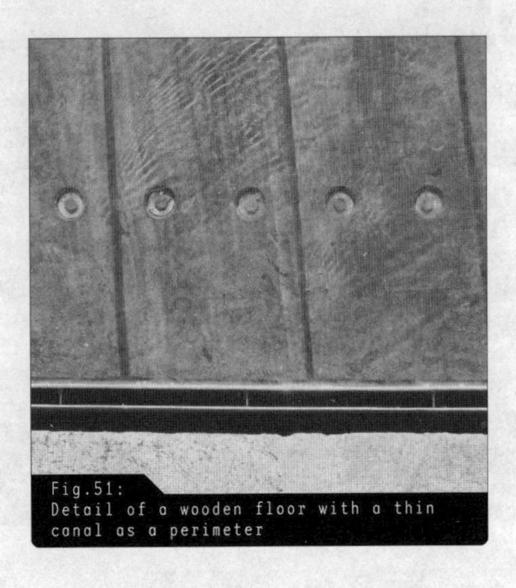

Fig.51:
Detail of a wooden floor with a thin canal as a perimeter

should be made using strong timber that is suitable for water construction (oak, larch, and tropical woods). › Figs 50 and 51

Walled basins

Walled basins of brink or natural stone are popular in climates with mild winters. Walled constructions offer a great deal of freedom to develop form, mix materials, and design fine-structured surfaces. › Fig. 52

However, a stable substructure is essential to prevent cracks caused by movement, or to be able to use frost-resistant bricks with low porous volume to protect against cracks in freezing weather. The many joints will always present a liner problem, which can be rectified by installing a prefabricated concrete basin. This will however affect the appearance of the design. Walled constructions are recommended for small architectural ponds and channels.

Natural stone

Without a doubt, natural stone is the most important and impressive material for water basins, fountain stones, shells, and troughs.

One simple variation would be a supporting concrete foundation combined with a thin upper layer of natural stone. Water channels are made of curbstones, cobblestones, or rough gravel. To save material and cost, smaller fountain blocks can be faced with thin sheets of stone; yet even precise joint work and caulking does not counteract the two-dimensional, rather fragmented quality of the stone-facing method. › Fig. 53

Fig.52:
Spring basin with thin, raw quarried natural stone, laid sandstone, and surrounding vegetation.

To avoid this problem, sandstone can be used to construct fountain blocks. This is a natural stone that is usually hand-processed by stonemasons. Solid, natural stone allows for a deeper sculptural treatment of the stones, which is necessary for curved and bowl-like forms, or decorative features. › Fig. 54 The joints between the stones will close and become waterproof if the stonework is executed with skill and absolute precision. Lead grouting is the most permanent method of sealing the joints.

Monolithic compositions are impressive when the uninterrupted force of stone interacts with the agility of water. › Fig. 55 Monolithic designs are contingent upon the pre-existing size of the blocks of stone, in other words, the thickness of the stone layer in the quarry, the extent to which the stone can to be processed, and transportation restrictions.

Natural stone can adapt to the slightest tolerances, so that controlling a precise flow in cascades or falls is possible even with low volumes of water. Processing the surface also enhances the overall look of the design. Ground or polished surfaces emphasize the shimmering quality of water, and accentuate form and elegance. Mottled or slightly sandblasted processes give the surface a velvety matt look that eventually weathers and becomes more interesting with age. Coarser processes such as stabbing or granulating not only give the stone a rustic appearance, but – depending upon the depth of the water – may also slow the water's flow and allow more moss to grow on the surface.

›✎

Fig.53:
A fountain block clad with thin plates of natural stone

Fig.54:
A fountain assembled from several solid blocks of sandstone

Unusual basins

There is a great variety of material available for smaller installations, for objects such as fountains and troughs where the liner and form are one and the same, perhaps decoratively designed and therefore a significant part of the concept. Regional context, locally available raw materials or traditional production or work methods can be expressed in this way. Glass, terracotta, or wax › Fig. 56 can be used, as well as aluminum, cast iron, bronze, lead, and other metals and alloys.

\\Tip:
Natural stone is subject to natural process of erosion. This is accelerated by existing pores, fine cracks, water penetrating the stone, cracks caused by frost wedging, and by tension caused by oscillating temperatures and after time, destroys the stone. To avoid this, only use natural stone with a good frost and water resilience that is manufacturer certified and guaranteed.

Fig.55:
A thin film of water flows along a monolithic natural stone fountain

Fig.56:
A rare example of a wax fountain in a temporary garden

DESIGNING THE PERIMETER

Designing the water's perimeter is one of the most essential phases of creating a water installation. How it is chosen and developed clarifies the design's overall coherence. Its form – an obstructing element with vegetation or walls, or an inviting one with footbridges, stepping stones, or steps – influences the way the water installation will be used and experienced. The perimeter defines the overall character, whether natural-looking or architectural, solid or finely structured.

Location

The perimeter is usually the most sensitive area of a water installation because it is the edge of the liner layer and the transition from water to surrounding environment. For one thing, water attracts the visitor and should be experienced as closely as possible, meaning there is more impact on the perimeters – so they need to be designed and built with extreme care. For another, perimeters are constantly exposed to fluctuating water levels, moisture penetration from the soil, or a steady swell of waves, all of which need to be counteracted by detailed planning and structural precision.

Perimeters have to be consistent and at least as high as the planned maximum water level as determined by inflow and outflow, and must be perfectly fitted and integrated into the form of the liner layer. Fluctuating water levels caused by evaporation or use must also be taken into account.

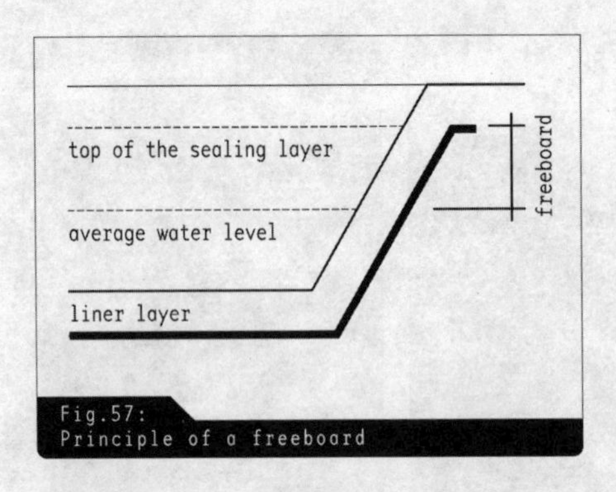

Fig.57:
Principle of a freeboard

Capillary action

Most materials have their own capillary action, a candlewick-like effect that is determined by the inner pore structure of the material. The capillary effect makes water sink into the soil beyond the visible water surface, which leads to water loss in basins, permanently waterlogged perimeters, and, depending on the material used, a decrease in the stability of the perimeter. >✎

Freeboard

A freeboard is commonly used in perimeter design to guard against damage and water loss. This is a securing continuation of the perimeter with a liner that extends above the maximum planned water level. › Fig. 57 The freeboard's height is determined by the size and momentum of the water installation, and its exposure and use. This is less necessary for smaller features such as small water bowls or birdbaths, where moving water splashed over the sides is not a problem. With mid-sized park ponds, a 20–40 cm freeboard is recommended. For natural bodies of water, the freeboard can extend to over 1 m.

✎

\\Tip:
Integrating a capillary barrier in the perimeter design will interrupt the capillary effect and thus guard against water loss and waterlogging at the edge zones. Suitable options include concrete or closely laid natural stone perimeters with low pore volume, as well as loose fills of coarse gravel or gravel sand with a low fine fraction content and large pore volume.

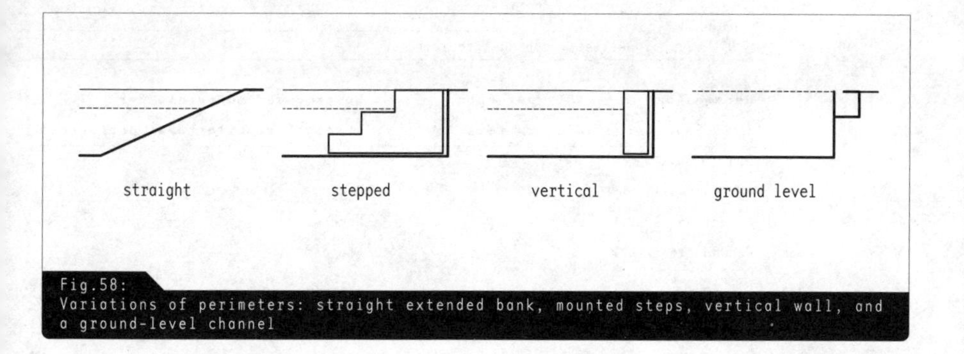

Fig.58:
Variations of perimeters: straight extended bank, mounted steps, vertical wall, and a ground-level channel

A freeboard gives the perimeter a more robust appearance and creates a division between the water experience and the visitor. The water table seems too "deep" and too small in relation with the overall proportions. This impression can be counteracted by the appropriate landscaping with vegetation and a more level, yet barely visible warping of the liner layer. Adding steps that lead into the water can even emphasize this technically produced high jump.

Bank

Perimeters are developed from the liner material, and by mixing elements, materials, and methods to form a continuous, formally cohesive, and durable liner. › Fig. 58 Perimeters are vulnerable to water movement, which can cause washouts and erosion, and strain due to warming or vibrations from heavy use can produce cracks around fixtures (such as footbridges) or at the seams where the material changes.

A simple way of forming perimeters is to slowly pull up the liner (for instance, mastic asphalt or mud liners) to reach the top of the freeboard. Here, the material's possible maximum slope as well as its minimum level of stability when wet need to be considered. This method creates relatively broad banks with long, shallow water zones, but stable structures like concrete walls or packed stone allow for steeper slopes and narrower banks of deeper water.

The choice of material to be used and the course of the perimeter influence the character of an edge zone as open, landscape-oriented, or straight and architectural. › Tab. 3

Natural-looking, landscaped perimeters

Naturally designed banks have a fluctuating interplay of water depth, perimeter width, materials, and plants and therefore need a perimeter that is varied and diverse. The irregular appearance allows a high level of

Tab.3:
Overview of perimeter coping materials

Recommended materials for open, landscape-oriented perimeters	**Recommended materials for straight architectural perimeters**
Grass	Walls
Hedges/bushes	Stairs
Gravel and sand	Channels
Packed stone/rock	Gabions
Wicker	

flexibility when choosing the type of liner › Chapter Technical parameters, Liners and basins the edge zone's form, the subtle integration of a freeboard, or when creating the correct habitat for the perimeter's vegetation. › Figs 59, 60 and Chapter Technical parameters, Plants

Structural, architectural perimeters

Hard-edged, architectural perimeters such as walls or fountain basins call for precise planning and construction, particularly with ground-level water surfaces, if the desired effect is to be visible. This method requires materials with a low dimensional tolerance (e.g. natural stone or concrete components), elements for precise water feed and drainage, and supporting technology (e.g. pump circulation). › Figs 61 and 62

WATER INFLOW AND OUTFLOW

Each system requires water inflow, for the day-to-day maintenance of the water level, and a water outflow, as well as a bottom outlet, if possible, to completely drain the water installation.

\\Tip:
The visible surface quality of the water can be controlled by the type of inflow. A method that employs low surface tension feed lines, located deep below the water's surface and above a basin connected upstream, produces a clear, smooth surface. Channels flush with the water's edge, which have a moderate surface tension feed line, make band-like waves and circular waves with a concentric source.

Fig.59:
A grass bank for a low current stream

Fig.60:
A straight, shallow extending gravel bank with scattered heavy rocks

Fig.61:
Hardwearing, curving perimeter made of in situ concrete

Fig.62:
Detail of a ground-level reflecting pool with a finely structured steel edging

Inflow and outflow are the starting and end point of any water design and have to be developed along with the overall concept. Their hydraulic capacity must be made to correspond with the system as a whole, and to operate either independently without the need for a pressurized system, or with a pressurized system using pumps and elevated tanks. These can be completely concealed or clearly visible, and should remain accessible for maintenance work and protected from harmful impurities by a filter system.

Concealed inflow and outflow

Concealed systems consist of a pressure hose and a suction hose, which are integrated in a pump system and hidden under the water surface. An additional water basin placed in front of the inflow can slow the speed of water flow in order to control, reduce, or avoid visible swirling on the water's surface. › Fig. 63 With depressurized systems, inflows and outflows can be concealed outside of the water using chutes, channels, and inflows.

› 

Fig.63:
Still water surface with a barely visible inflow built into the basin floor

Fig.64:
Decorative fountain pipe

Fig.65:
Water source made of fired clay

Fig.66:
Protective grating over a pond spillover in the form of a leaf

Visible inflow and outflow

The possibilities of a visible, and therefore, designable inflow and outflow system are very diverse. In simple cases, they are designed as a simple water figure rising from the surface, perhaps as a constant gurgle or pulsing geyser. Technical constructions, such as the overflow from dams, coastal pumping stations, or pumps that have a rough technical charm can be designed and integrated in a manner that suits this context. One simple example would be extended water outflow pipes › Fig. 64 made of cast iron or bronze, decorated or plain, straight or curved. Water can gush from built elements, natural rock, or processed stones, bowls, chutes, or steel grating; it can surge over or vanish into them. Finally, the water sources can be collected and designed to enhance this effect, by using sculptural forms, artificial sculpturally emphasized elements, allegorical

Fig.67:
Outlet in the middle of a reflective pool

figures, fairytale figures, or fountain saints that all spray, squirt, spill and catch water. › **Figs 65, 66 and 67**

Bottom outlet

For maintenance work and emptying the pool in winter, › **Chapter Technical parameters, Winter protection** the bottom outlet should be designed as a simple technical element that is barely visible but easy to operate. It can be an outflow unit located at the deepest part of the basin floor that can be opened and closed by a sliding valve, that is, a bilge pump that can also be a part of the entire pump system, or, in smaller installations, a simple standpipe for overflow, which can be completely removed when emptying the basin.

MOVEMENT

Movement influences the visible character of water. Installations with moving water can be classified in three ways: flowing, falling, and rising.

\\Note:
In connection with the inflow and outflow, quieter water movements and technologically simple solutions can, for the most part, be developed as depressurized along with the natural flow of the water. Lively, swirling water surfaces, in contrast, need enough pressure to propel the water and, hence, the necessary technology.

Fig.68: Moving water surface in a sculpturally designed water channel

Fig.69: A dam creates some turbulence and different speeds of flow

Flowing water

Bodies of flowing water like brooks or channels play with water flowing downward from a high position to a lower one. They can be simple channels, a sequence of basins, or a combination of natural or artificial, linear or meandering forms. Channels require permanent water feed and will run dry if this supply is interrupted. It is wise to retain some water with subtle, built-in barriers and hollows in order to keep it active when the system is turned off or if there is a prolonged dry spell.

The amount of water required for an installation can be calculated, using the diameter of the outflow point, the slope, the degree of surface roughness in the channel, and the speed of flow. With uniform slopes, widening the section slows the flow. Narrowing it increases the speed. › Fig. 68 The appearance of waves and whirlpools on the water's surface is influenced by the course of the channels and how frequently it shifts direction, by the speed of flow, existing swells and breaks, and by the surface structure of the section. › Fig. 69

›✎

Falling water

Waterfall installations work with free falling volumes of water. For this method, water is collected before an overhang, slowed and retained before flowing over the projection and falling over the edge into water below.

The overhang, edge, and speed of flow determine the water cascade's clarity, constancy, or calmness. › Fig. 70 The more calm, clear, and consistent

Fig.70:
A film of water that increasingly fragments

a waterfall is intended to be, the more precisely the constant flow of water will have to be calculated, and the more accurately the surface, construction, and horizontal installation of the overhangs and edges will need to be developed. Natural, "wild" waterfalls have moving, turbulent sheets of water and misty spumes. They function with irregular edges, changing currents, and impact stones to enhance the noisy effect. Since it is difficult to predetermine the actual optical effect of falling water, it is best to test-run the installation during the construction phase to make any necessary improvements or adjustments.

\\Tip:
If a powerful waterfall lands in a still pool of water, the reverberations on the water's surface will continue far out into the water. In order to keep the calm character of the still water, a stone can be placed in the stream, just behind the impact from the falls. This interrupts the visible movement on the water and calms the surface.

Waterfalls, particularly those with precise, clear sheets of water, react sensitively to debris. Even slight water impurities such as leaves are enough to tear the film of water, making its fall irregular. A good location, screens or filters located upstream from the overhang can guard against this problem and keep the flowing water clear and free of impurities.

Rising water

Fountain installations consist of rising jets of water of varying heights and strengths. They can be arranged alone or in groups, and be constant or rhythmic in height and movement. Their individual form results from the specific combination, water pressure, and shape of the fountain nozzles and fixtures. › Fig. 71

Single jet nozzles › Fig. 72 create a clear, wind-resistant, full water jet that can rise as high as 14 m and be pitched at a 15° angle from the vertical. A multi-jet nozzle creates jets of water that fall diverging or merging together. When mounted on a revolving base, the recoiling water creates a rotating, screw-shaped form. Water-saving hollow jet nozzles are used for fountains with an elevation of up to 80 m. Water film nozzles create closed, but wind-sensitive forms like dome-shaped water bells. Fan nozzles produce the impressive, full picture of a large, closed, fan-shaped water spray at a 30° angle. Finger nozzles produce a vertical or diagonal, bizarre, broken sheet of water. If air is added to the nozzles, the spray becomes full of contrast, bubbly, and foamy, with a powerful wind-resistant body. › Fig. 81 Elaborate switching mechanisms and fast pressure valves produce swaying, dancing water shapes.

Mist fountains in outdoor installations are a special form of fountain installations, where water is finely sprayed using high-pressure valves. The fine mist lowers the air temperature, which is pleasantly fresh in summer. It produces constantly changing, fleeting shapes that are playfully

\\ Tip:
If the rate of water flow is low, adhesion power will stop a clear stream from forming, and even the transition into a free fall. The water gets "stuck" in the fall, wanders back, or falls in haphazard, broken streams. Carving a draining lip on the underside directly after the edge, for instance in the form of a milled groove in a stone, can avoid this problem.

\\ Note:
It is important to consider the correct size of the basin when designing a fountain, so that water carried by the wind can be collected. The recommended ratio between the height of the stream and the width of the basin is 1:2, and 1:3 in case of strong wind. In tight spaces, this can lead to difficulties with adjacent uses. This can be rectified by correct positioning together with an automatic wind gauge control system.

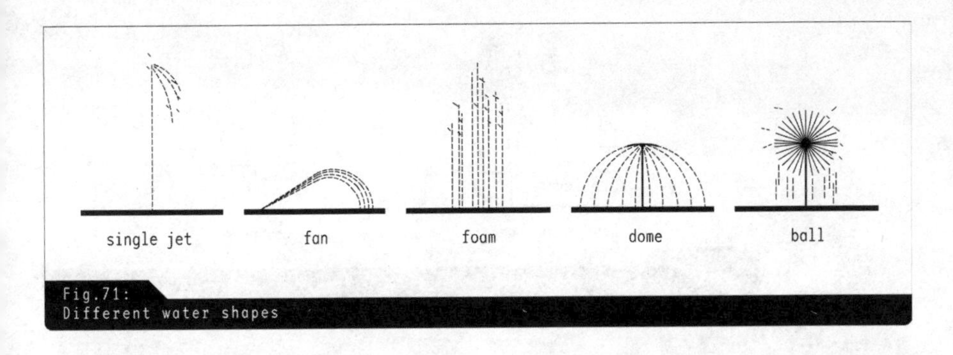

Fig.71:
Different water shapes

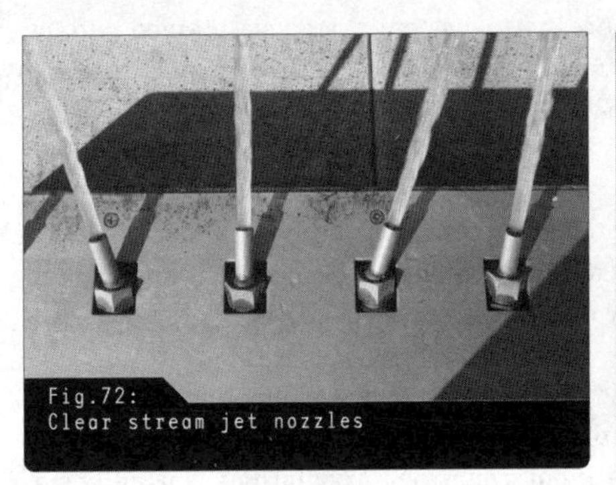
Fig.72:
Clear stream jet nozzles

Fig.73:
Aquatic recreation area with temporary banks of mist

enchanting and fluctuate between transparent and opaque. The installations' fine nozzles make them sensitive to impurities and calciferous water, and their level of water loss depends on wind conditions, and they therefore need a constant water inflow. › Fig. 73

PUMPS AND TECHNOLOGY

The movement of water is the result of balancing differences in height or pressure that is artificially created by pumps, pipe systems, and reservoir tanks.

Pump types

Submersible pumps are placed within the body of water in the middle of the basin to save space, or protected inside a nearby shed. Anchored to a platform, they can float on the surface in the middle of a lake and are used in the design of large fountain installations.

Dry pumps remain outside the water in a separate pump chamber and are connected to a reservoir by cables. Dry pumps are more complex

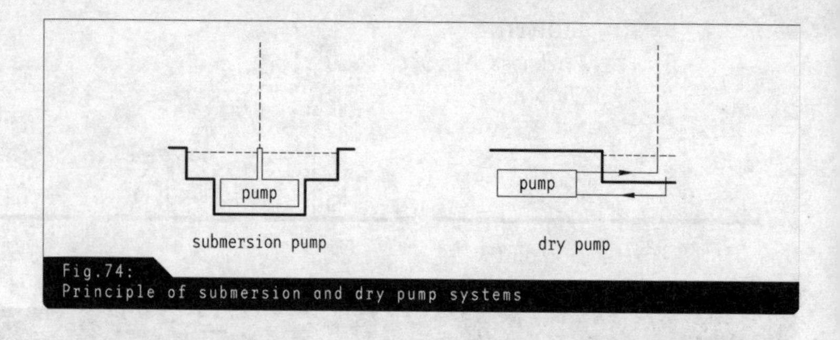

Fig.74:
Principle of submersion and dry pump systems

to integrate and therefore more expensive, but can be maintained without great effort, making them more cost-efficient. They are common in large and public installations. › Fig. 74

Installation

Pumps should be enclosed with a protective grille to guard against accidents and should be protected against impurities with a filter and a grit trap. They are operated automatically by a float switch or magnetic switch control and computer programs.

The type and size of the installation and the planned movement and volume of water determine the choice of pump. For smaller, simpler installations, the wide availability of prefabricated solutions is usually sufficient. Pumps greatly affect the running costs of an installation, which makes it advisable to seek advice from specialist engineers and the manufacturer when considering the design. › ›

\\Note:
In waterfalls or running water installations without their own reservoirs, pumps at rest collect the otherwise moving water in the last, bottom compartment. The required volumes can be stored in visible ponds or basins, which can be seen in clear fluctuations of water levels, and also require the appropriate dimension and perimeter. Concealed, underground collecting and compensating tanks reduce the visible effect and are therefore preferred for design reasons.

\\Example:
Historic installations, such as the animated fountains in Kassel-Wilhelmshöhe's landscape park, Germany, often function using elevated tanks that are kept full by a low, but steady level of pump power, and then emptied as a spectacular event over a specified period of time. With this technique, animated fountains are not in operation at any other time, but are limited to predetermined intervals. In modern installations, pumps are developed for the specific object and allow flexibility, control, and operation round the clock.

LIGHTING

Darkness robs water of its optical power of attraction, quickly turning it into a dark, impenetrable surface. A lighting concept specifically created for the design can animate the water at night and dramatize it with specific lighting.

Outside light sources

Light hitting the surface of water from the outside creates a soft reflection on the otherwise dark plane. The intensity of the reflection depends on the light source's brightness, the distance between the light and the water surface, the color spectrum, and movement on the surface of the water. Colors can also be added by the light source, but they look paler and less saturated when reflected.

Underwater light source

Intense luminosity can be achieved by placing light sources inside the water. They can be located along the basin perimeter, in the soil, or close to fountains. Depending on the brightness of the lighting, the distance to the surrounding fixtures, and the water's transparency, reflections on the basin walls can illuminate the entire body of water. The different angles of refraction emanating from an illuminated body of water with movement on the surface produce a constantly changing structured pattern, and project a play of light and shadow onto the surrounding environment.

Only specific underwater lights or cold lights can be used for underwater lighting concepts. In this procedure, light from an outside projector is directed under the water via glass fibers. Color can be added by using color filters.

PLANTS

Images of water almost always include plants: the water lilies in Monet's garden, a picturesque willow at a pond's edge, the soft, swishing sound of reeds lining a beach. Even architectural designs are willing to interrupt their strict linearity and precise forms by integrating the soft contours and movements of plants.

Their height, density, colors, and leaf structure change during the course of the year. While trees can display their impressive charm in large spaces like parks or landscaped grounds, smaller spaces are characterized in detail mainly by shrubbery.

Choosing a location

Plants will only thrive in a location that suits their individual needs. These habitats are determined by a plant variety's specific needs for light, soil, and water. Trees such as willows or alders are more tolerant than

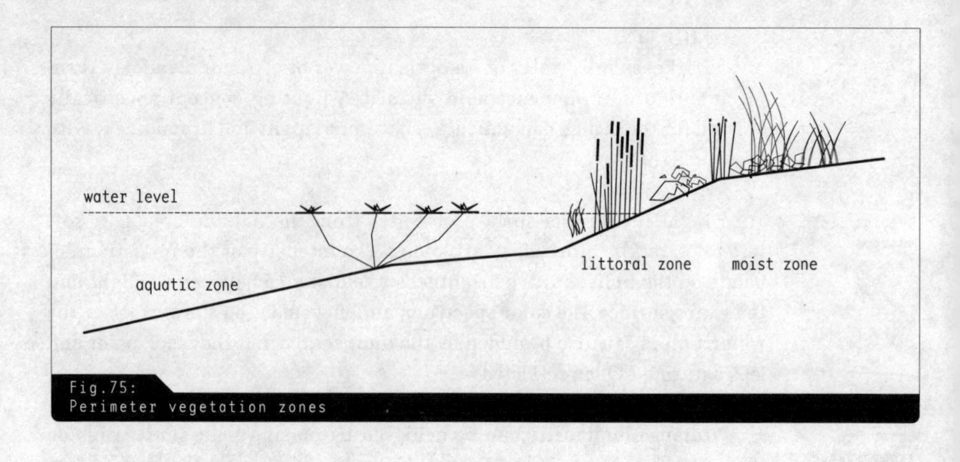

Fig.75:
Perimeter vegetation zones

shrubs, which respond sensitively to changes and fluctuations in the environment.

The habitats of plants can be allocated to different zones according to their respective ideal distance to water. › Fig. 75 Moist zones are located just beyond the water's edge. The capillary effect keeps the ground constantly wet, and occasional flooding is possible. This is a perfect habitat for floral shrubbery and marsh plants, which can tolerate constant moisture. Semi-aquatic plants may grow in water of up to 60 cm deep, depending on their variety. Deeper areas are ideal for plants that are rooted in the soil but have leaves that float on the water's surface; or for floating aquatic plants and underwater plants.

Plants are sensitive to water movement, which can restrict the choice of plant variety, according to the strength of the current. Blades from cane break will bend and underwater plants will be uprooted if the movement of the water is too strong. Water that drops onto the leaves, for example from an animated fountain, can have a magnifying glass effect and produce unsightly burns on the plant.

Growing power

Given the right habitat with the correct nutrients and amount of water, aquatic plants thrive with astounding vitality and growth, changing the appearance of the water design in a short period. Some plant varieties will be suppressed; water surfaces can become completely overgrown. Medium-term plant stability can only be assured if the size of the overall installation is large enough to accommodate the planted area, if there is a nutrient-poor plant substratum, if the vegetation is planted in baskets

Fig.76:
Willows are a symbol for water and riverbanks in every culture

Fig.77:
Water lilies in sunken baskets placed in a geometric reflecting pool

that restrict its growth, and if the installation is regularly maintained.
› Fig. 77

The right selection and combination of plants requires moderation in variety and density, great care and experience, if an attractive, stable, and visible structure is desired.

›  › 

WATER QUALITY

Despite the increasing number of aesthetically demanding examples that focus thematically on pollution, change, and transformation, there are other cases that present the desired ideal of clear, fresh, and transparent water. Murky water, algae, or even brackish, foul-smelling bodies of water

\\Tip:
Biological factors affect the natural presence of plants near water. This has influenced our visual experience so much that those plants we often see near water, such as willows of reeds, have even come to symbolize water. These may be used strategically as design elements (see Fig. 76).

\\Note:
Tables 1–3 in the Appendix provide a basic guide to varieties of water plants. Additional information and design approaches can be found in *Basics Designing with Plants* by Regine Ellen Wöhrle and Hans-Jörg Wöhrle, Birkhäuser Verlag, Basel 2008, as well as in *Perennials and their garden habitats* by Richard Hansen and Friedrich Stahl, Cambridge 1993.

Fig. 78:
Thick algae cover the entire surface of the water

demonstrate not only a lack of biological equilibrium, but also a failed design. › Fig. 78

Biological equilibrium

Biological equilibrium is a prerequisite for clear, transparent water. It can be achieved only if substrate producers and substrate eaters – plants, animals, and microorganisms – are in proportion to one another. This can be achieved if the water is slightly shaded, if there is water movement that brings about oxygenation, if the body of water is larger, deeper and hence cooler, is not heavily exploited, and has a low animal population.

In closed systems, biological equilibrium requires the water to be changed or cleaned continually by both natural and artificial means, that is electronic or chemical processes.

Natural cleaning

Larger, natural installations, such as swimming ponds with a balanced proportion of exposed water surface and perimeters with a self-cleaning capacity, almost exclusively use a natural cleaning method. This cleans the water within a constructed wetland that contains reeds, and sometimes cattail, mud rush, or sedges. The plants' roots are a perfect habitat for microorganisms that break down foreign nutrients and harmful substances, supply oxygen to the water, and contribute to cleaning it.

Hybrid systems that support the natural cleaning capacity by mechanically recirculating the water and using sand filters along the edge zones are recommended for installations that are heavily used, such as public swimming ponds.

Artificial cleaning

Artificial cleaning simulates the natural processes but is concentrated on a small space. In the first mechanical-hydraulic phase, large debris is removed from the water by screens, and fine floating sediments by skimmers, sieves, and crystal-quartz sand filters. In the following chemical phase, the pH of the water is neutralized by adding acidity or leaching with an automatic pH regulator. Bacteria are kept at bay with bactericidal products such as chlorine, which is controversial when used outdoors, or by adding hydrogen peroxide or silver oxide.

Complete technical conditioning is common for architectural installations without vegetation, which except for freshwater, have no natural cleaning capacity. Installations such as aquatic recreation areas that are heavily used require hygienic, high-quality water treatment.

SAFETY

Water has always been both fascinating and threatening. The safety requirements when using water as a design element are therefore very high, to guard against the risk of accidents. The different possible dangers include deep water, entering the water unsupervised, and the underwater technical equipment such as pumps or inlet pipes.

Precautions

Having to integrate safety measures after the installation has been constructed is almost always detrimental to the design. It makes more sense to address these issues during the design phase and to consider the aspect of safety as a part of the overall concept. The most important points to consider include:

_ Controlled access to the water: Closed and accessible banks should be designed clearly so that visitors can easily differentiate between the two. Accessible banks should allow the water to be accessed by ramps, steps, stepping-stones, or handrails.
_ The right water depth: A safe water depth is essential in accessible areas where children play. In water that is already deep, an underwater grating can be installed to create a safe, relatively shallow depth. › Fig. 79

Fig.79:
A safety grating installed just under the surface of a deeper body of water

- Good escape routes out of the water: If there is an accident, it should be easy to leave the water without requiring help from others. This can be achieved by a relatively gentle slope of 1/3 or less, or with steps or ladders.
- Equipment safety: Technical equipment such as inlet pipes or immersion pumps should be covered with grating to guard against accidents.

WINTER PROTECTION

Depending on the region in question, the cold season, when water freezes and develops a particular aesthetic effect as snow and ice, can encompass up to half the year. Due to the anomaly of water, that is, its capacity to expand up to 10% in volume when frozen, low temperatures pose a threat to a water installation's technical equipment. To protect it, most installations are emptied and closed at the beginning of winter. They look abandoned, gloomy, and unsightly compared to permanent installations. A good design always considers the installation's year-round appearance, taking into account the closed phase (such as winter) from a design and a technical point of view, and developing a suitable design response.

Technical delays

Still water freezes more quickly than moving water. Heating the water and basin, or adding chemicals, can prevent or at least delay the water from freezing, but the high construction and maintenance costs make this solution less attractive.

Emptied installations

Installations at risk of frost are usually closed and completely emptied at the beginning of the frost period. During this period, an outlet should be kept open to drain water that accumulates from precipitation, seepage, and condensation, which can also freeze.

Simple constructions can be closed during winter and do not need to be protected. Valuable structures such as fountains with sensitive sculptures and decorative elements are protected mostly by wood casings.

Filled installations

Installations with liners that are sensitive to dryness, such as clay or silt, or that have a dense flora and fauna population remain filled with water during the winter season. There needs to be enough space at the perimeter to allow for the expansion of ice so as to avoid damage to these installations, and the existing closed, water-filled pipes have to be protected against frost. How deeply frost penetrates the soil depends on the local climate. The average frost depth ranges from 0.8 to 1.2 m.

Water plants that have been correctly chosen for the location and its climate will not need additional winter protection. To ensure that the aquatic animals survive, there has to have been an adequate amount of frost-protected water to offer as much habitat as necessary below the ice layer.

Spring cleaning

The end of the winter season signals an intense period of cleaning, where the protective casing is removed, leaves, mud, and refuse are eliminated, and the luxuriant plants are returned. Before reopening the installation, the equipment is serviced, and the installation is checked for any possible damage.

COST EFFECTIVENESS

Of the three design elements of landscape architecture – vegetation, topography, and water – it is the last that has the greatest long-term influence on cost. The initial costs, however, are not the most significant. Mid-size, artificial water installations and simple constructions cost about the same to build as mounted ones. Even more elaborate animated fountains are relatively reasonable to produce because of their concentrated size.

It is the constant care, maintenance, and repairs that make a water installation costly in the medium term. The economic factor can hinder good designs from being realized, or could mean technically perfect, built installations being closed down because of a lack of finances.

Sustainable designs should therefore try to create a lasting and cohesive design, in terms of cost, during the planning and conception phase. This is possible if existing elements are exploited and designed optimally, and it becomes structurally feasible if the necessary technological standards are lowered, and if the workmanship is good down to the last detail.

Optimizing the concept

Conceptually, solutions that rely on the site and what is locally available are in a better position to make the best use of synergies and cost. Buried brooks can be rediscovered and opened, existing basins used in a new way, or rainwater allowed to playfully run through open channels, or be directed into cisterns and ponds. › Fig. 80

The appropriate standards

Greater tolerance in the design and a slight lowering of standards allow water installations to be realized and maintained more easily. This allows for a less expensive integration of natural processes, for example in the case of near-natural swimming ponds that rely on self-cleaning ability, rather than traditional swimming pools that need elaborate and costly technical installations. ›

Optimizing the construction

In order to avoid the need for subsequent repairs, it is recommended to choose a durable solution over an apparently cost-effective construction. Any possibility of simplifying and reducing repairs and maintenance should be exploited; for example, protecting against excessive amounts of leaves by providing easy accessibility for maintenance, or by calibrating the installation with others belonging to the same operator in order to exploit operational synergies and the availability of spare parts.

\\ Example:
In 2002 in Grossenhain, Germany, the outdoor swimming pool was replaced by a natural swimming pool. 2000 m^2 of regeneration surface is adjacent to 3000 m^2 of swimming pond, which is used by up to 1,800 visitors a day. The natural swimming pool reduced operational costs by 40% compared to conventional swimming pools.

Fig.80:
A water basin that collects rainwater and has vegetation and a wooden deck form the focal point of a public square

Fig.81:
Water Block in front of the Louvre

With artificial installations, it is important to note:

_ that the water basin should be as shallow as possible so as to minimize the cost of filling and replacing water;
_ that pumps should be installed individually according to the correct requirements in order to optimize energy and water use;
_ that animated fountains should not be located in windy passages so as to avoid excessive water loss;
_ that thorough winter protection measures will guard against damage to technical equipment.

With "natural" installations

_ the body of water should be as large as possible so as to avoid warming too quickly;
_ vegetation should only be planted in baskets to contain their growth;
_ fish should not be introduced because the nutrients they bring encourage the growth of algae.

IN CONCLUSION

Water is an essential, almost unavoidable, element when designing outdoor spaces. Its symbolic character, vivacity, its unique emotional quality, and immanent diversity provide the opportunity to develop a suitable but distinctive overall concept with visual poetry. Water designs grab people's attention – how it functions naturally, how one feels it work, and how it moves the observer's emotions. Contact with water triggers a deeply rooted reflection in relation to the overall context, a response to the uniqueness of the site, accuracy of material, and quality of workmanship as a basis for a design's success.

Water is freedom. Within the boundaries of the physical specifications, there is an almost infinite amount of room for creativity. Water generates curiosity and the desire to experiment. It offers long-lasting appeal and is always surprising. This text aims to be an introduction – an incentive to observe, a call to experiment, and an encouragement to explore the path less trodden.

APPENDIX

AQUATIC PLANTS

The table below provides an overview of suitable aquatic plants and can serve as an introductory guideline to planting vegetation.

Tab.1:
Selection of shrubs for moist zones

Variety	English name	Exposure to sun	Height	Bloom time (month)	Bloom color
Ajuga reptans	Common bugleweed	Half-shade	10–20 cm	IV–V	Blue
Caltha palustris	Marsh marigold	Sun to half-shade	30 cm	III–IV	Golden yellow
Cardamine pratensis	Cuckoo flower	Half-shade to shade	30 cm	IV–VI	Lilac-pink
Carex elata "Bowles Golden"	Bowles' golden sedge	Sun to half-shade	60 cm	VI	Brown
Carex grayi	Grey's sedge	Sun to shade	30–60 cm	VII–VIII	Green-brown
Darmera peltata	Indian rhubarb	Sun to half-shade	80 cm	IV–V	Pink
Eupatorium species	Gravel root	Sun to shade	80–200 cm	VII–VIII	Dark pink
Filipendula ulmaria	Meadowsweet	Sun to half-shade	120 cm	VI–VIII	Cream
Frittilaria meleagris	Checkered lily	Half-shade	20–30 cm	IV–V	Violet
Geum rivale	Water avens	Sun to half-shade	30 cm	IV–V	Golden brown
Gunnera tinctoria	Giant rhubarb	Sun	150 cm	IX	Red
Hemerocallis wild species	Daylily	Sun-half-shade	40–90 cm	V–VIII	Yellow/ orange
Iris sibirica	Siberian iris	Sun to half-shade	40–90 cm	V–VI	Violet
Leucojum vernum	Spring snowflake	Half-shade to sun	20–40 cm	II–IV	White
Ligularia species and hybrids	Ragwort	Sun to half-shade	80–200 cm	VI–IX	Yellow/ orange
Lysimachia clethroides	Gooseneck loose-strife	Sun to half-shade	70–100 cm	VII–VIII	White
Lythrum salicaria	Purple loose-strife	Sun to half-shade	80–140 cm	VII–IX	Violet
Myosotis palustris	Water forget-me-not	Sun to half-shade	30 cm	V–IX	Blue
Primula species and hybrids	Primrose	Half-shade to shade	40–60 cm	II–VIII	White/ yellow/pink-red/ orange/ lilac-violet
Tradescantia-Andersoniana hybrids	Andersons Spiderwood	Sun	40–60 cm	VI–VIII	Blue/violet/ white/ carmine
Trollius europaeus/Trollius hybrids	Globe-flower	Sun to half-shade	40–70 cm	IV–VI	Yellow/ orange
Veronica longifolia	Garden speedwell	Sun	50–120 cm	VII–VIII	Blue

Tab.2:
Selection of shrubs for banks

Variety	English name	Exposure to sun	Water depth	Height	Bloom time	Bloom color
Acorus calamus	Sweet flag	Sun to half-shade	0–30 cm	to 120 cm	V–VI	Yellow
Alisma plantago-aquatica	Water plantain	Sun to half-shade	5–30 cm	40–80 cm	VI–VIII	White
Butomus umbellatus	Flowering rush	Sun to half-shade	10–40 cm	100 cm	VI–VIII	Pink
Calla palustris	Water arum	Sun to half-shade	0–15 cm	to 40 cm	V–VII	White-yellow
Caltha palustris	Marsh marigold	Sun to shade	0–30 cm	to 30 cm	IV–VI	Yellow
Carex pseudocyperus	Cypress-like sedge	Sun	0–20 cm	80 cm	VI–VII	–
Equisetum fluviatale	Swamp horsetail	Sun to half-shade	0–5 cm	20–150 cm	–	–
Hippuris vulgaris	Common marestail	Sun to half-shade	10–40 cm	40 cm	V–VIII	Green
Iris pseudocorus	Yellow flag	Sun to shade	0–30 cm	60–80 cm	V–VIII	Yellow
Phragmites australis "Variegatus"	Common reed (yellow striped variety)	Sun	0–20 cm	120–150 cm	VII–IX	Brownish-red
Pontederia cordata	Pickerel weed	Shade to half-shade	0–30 cm	50–60 cm	VI–VIII	Blue
Sagittaria sagittifolia	Arrowhead	Sun	10–40 cm	30–60 cm	VI–VIII	White
Scirpus lacustris	Bulrush	Sun to half-shade	10–60 cm	to 120 cm	VII–VIII	Brown
Sparganium erectum	Bur-reed	Sun to half-shade	0–30 cm	100 cm	VII–VIII	Green
Typha angustifolia	Narrow leaf cattail	Sun	0–50 cm	150–200 cm	VII–VIII	Reddish brown spadices
Typha minima	Miniature cattail	Sun	0–40 cm	50–60 cm	VI–VII	Brown bells
Veronica beccabunga	Brooklime	Sun to half-shade	0–20 cm	40–60 cm	V–IX	Blue

Tab.3:
Selection of aquatic plants

Variety	English name	Exposure to sun	Height	Bloom time	Bloom color
Callitriche palustris	Common water-wort	Sun to half-shade	10–30 cm	–	–
Ceratophyllum demersum	Hornwort	Sun to half-shade	30–100 cm	Summer	Not visible
Elodea canadensis	Canadian pond-weed	Sun	20–100 cm	–	–
Hydrocharis morsus-ranae	European frog-bit	Sun to half-shade	from 20 cm	VI–VIII	White
Myriophyllum verticillatum	Myriad leaf	Sun to half-shade	30 cm	VI–VIII	Pink
Lemna minor	Duckweed	Sun	from 20 cm	–	–
Lemna trisulca	Star duckweed	Sun	from 20 cm	–	–
Nuphar lutea	Yellow water-lily	Sun to shade	40–100 cm	VI–VIII	Yellow
Nymphaea alba	White water-lily	Sun	60–100 cm	V–VIII	White
Nymphaea-Hybriden, e.g. "Anna Epple", "Charles de Meurville", "Marliacea Albida", "Maurica Laydeker"	Water-lilies: hybrids	Sun	40–60 cm	VI–IX	Pink/ burgundy/ pure white/ purple
Nymphaea odorata "Rosennymphe"	Fragrant water-lily	Sun	40–60 cm	VI–IX	Pink
Nymphaea tuberosa "Pöstlingberg"	Tuberous water-lily	Sun	60–80 cm	VI–IX	White
Nymphoides peltata	Yellow floating-heart	Sun to half-shade	40–50 cm	VII–VIII	Yellow
Potamogeton natans	Broad-leafed pondweed	Sun to half-shade	40–100 cm	VI–IX	White
Ranunculus aquatilis	Water crowfoot	Sun	30 cm	VI–VIII	White
Stratiotes aloides	Water soldier	Sun to half-shade	30–100 cm	VI–VIII	White
Trapa natans	Water chestnut	Sun	40–120 cm	VI–VII	Light blue
Utricularia vulgaris	Bladderwort	Sun to half-shade	20–40 cm	VII–VIII	Yellow

LITERATURE

Alejandro Bahamón: *Water Features*, Loft Publications, Barcelona 2006

David Bennett: *Concrete*, Birkhäuser Verlag, Basel 2001

Bert Bielefeld, Sebastian El khouli: *Basics Design Ideas*, Birkhäuser Verlag, Basel 2007

Ulrike Brandi, Christoph Geissmar-Brandi: *Lightbook, The Practice of Lighting Design*, Birkhäuser Verlag, Basel/Boston 2001

Francis Ching: *Architecture: Form, Space and Order*, Van Nostrand Reinhold, London/New York 1979

Christian Cajus Lorenz Hirschfeld: *Theory of Garden Art*, University of Pennsylvania Press, Philadelphia, PA 2001

Paul Cooper: *The New Tech Garden*, Octopus, London 2001

Herbert Dreiseitl, Dieter Grau, Karl Ludwig: *New Waterscapes*, Birkhäuser Verlag, Basel 2005

Edition Topos: *Water, Designing with Water, From Promenades to Water Features*, Birkhäuser Verlag, Basel 2003

Renata Giovanardi: *Carlo Scarpa e l'aqua*, Cicero editore, Venice 2006

Richard Hansen, Friedrich Stahl: *Perennials and Their Garden Habitats*, Cambridge University Press, Cambridge 1993

Teiji Itoh: *The Japanese Gardens*, Yale University Press, New Haven/London 1972

Hans Loidl, Stefan Bernhard: *Opening Spaces*, Birkhäuser Verlag, Basel 2003

Gilly Love: *Water in the Garden*, Aquamarine, London 2001

Ernst Neufert: *Architects' Data*, Blackwell Science Publishers, Malden, MA 2000

OASE Fountain Technologie 2007/08, http://www.oase-livingwater.com/

George Plumptre: *The Water Garden*, Thames & Hudson, London 2003

TOPOS 59 – Water, Design and Management, Callwey Verlag, Munich 2007

Udo Weilacher: *Syntax of Landscape*, Birkhäuser Verlag, Basel 2007

Stephen Woodhams: *Portfolio of Contemporary Gardens*, Quadrille, London 1999

PICTURE CREDITS

All drawings plus the illustration on page 8 (Latona Fountain in Versailles, LeNotre), 2, 3 (Bundesplatz in Bern, Staufenegger + Stutz), 6, 7, 8 (City park in Weingarten, lohrer.hochrein with Bürhaus), 10 (Landscape Park in Basedow, Peter Joseph Lenné), 11 (Park Vaux-le-Vicomte in Meaux, Le Notre), 12 (Fountain in Winterthur, Schneider-Hoppe/Judd), 13, 14 (Japanese Garden in Berlin, Marzahn, Shunmyo Masuno), 15 (Zeche Zollverein in Essen, Planergruppe Oberhausen), 16 (Villa Lante Garden in Viterbo, da Vignola), 18 (Spa Gardens in Bad Oeynhausen, agence ter), 19 (Bertel Thorvaldsens Plads in Copenhagen, Larsen/Capetillo), 20 (Fountain in Park Schloss Schlitz, Schott), 21 (Duisburg Nord landscape park, Latz and Partner), 22 (Spa Gardens in Bad Oeynhausen, agence ter), 23 (Cloister Volkenroda, gmp), 24 (Oberbaum City in Berlin, Lange), 25 ("Fuente" drinking fountain, Battle + Roig), 26, 27 (Allerpark in Wolfsburg, Kiefer), 28 (Botanical garden in Bordeaux, Mosbach), 29, 30 (Outdoor swimming pool in Biberstein, Schweingruber Zulauf), 31 (City park in Burghausen, Rehwald), 32 (City garden in Weingarten, lohrer.hochrein mit Bürhaus), 33, 34 (Paley Park in New York, Zion & Breen), 35 (Oriental Garden in Berlin, Louafi), 36, 37, 38 rechts (Mosaics in Saint-Paul de Vence, Braque), 39 (Villa d'Este in Tivoli, Ligori and Galvani), 43, 44, 45 (Cemetery in Küttigen, Schweingruber Zulauf), 46 (Promenade in Kreuzlingen, Bürgi), 47, 48 (Fountain in Dyck, rmp), 49 (Fountain in Bregenz, Rotzler Krebs Partner), 50 (Fountain in Valdendas), 51, 52 (Garden in Hestercombe, Jekyll), 54, 55 (Sundspromenaden in Malmö, Andersen), 56 (Garten der Gewalt (Garden of Violence) in Murten, Vogt), 59, 60 (Brückenpark in Müngsten, Atelier Loidl), 61 (Ankarparken in Malmö, Andersson), 62, 64, 65 (Fountain in Saint-Paul-de-Vence, Miro), 66 (City garden in Erfurt, Kinast, Vogt, Partner), 67, 68 (Villa Lante Garden in Viterbo, da Vignola), 69, 70, 72, 73 (Parc de la Villette in Paris, Tschumi), 76 (Chinese Garden in Berlin, Pekinger Institut für klassische Gartenarchitektur), 77 (Open spaces on Max-Bill-Platz in Zurich, asp), 78, 79, 80 (Inre Strak in Malmö, 13.3 landskapsarkitekter), 81 (Louvre in Paris, Pei):
lohrer.hochrein landschaftsarchitekten

Figure 40 (Elbauenpark Magdeburg, Ernst. Heckel. Lohrer with Schwarz und Manke), 53, 63 (Rubenowplatz in Greifswald, lohrer.hochrein):
Hans Wulf Kunze

THE AUTHOR

Axel Lohrer, Dipl.-Ing. (FH), practicing landscape architect and partner at lohrer.hochrein landschaftsarchitekten in Munich and Magdeburg.

P9

导言：水的多样性

景观设计是一个丰富而复杂的学科。它能够提取自然元素创造建筑形式和结构，同时还可以从充满内在神秘力量、拥有深层魅力的大自然中，吸取其无限性和多样性。

水在自然元素中具有独特的地位。人类与水的关系是复杂的、矛盾的，总在过多和过少之间纠结。水是生命的基础。它的能量、愈合功能、光和冥想的灵感吸引着我们所有人。但水还含有一种危险的元素，它可以令人恐惧和敬畏，干旱或洪涝也会威胁生命。

以水为设计元素进行的设计创作，总是会引发此领域的冲突，不过水也能够激发渴望、回忆以及技术可能性。

野性

水呈现出难以控制的特质，因此至纯、自由、力量无穷。它象征着一个被技术驾驭的世界的对立面。这些方面可以以咆哮的瀑布、欢快十足的喷泉或水雾弥漫的雕塑这些形式被动态地表现出来。

魔力

水既是无生命的东西又是生命的象征。它是神话和自然哲学的一个重要构成要素。在世界的许多地方，它都扮演着重要的角色，尤其是在水与生存问题联系在一起的地方。作为一种神奇的或者动态元素，水的形象出现在传说、歌曲或象征性事物之中，或在水景设计概念上被关注，或者成为雕塑的附加要素。

净化、放松

因为它在洗涤沐浴过程中扮演的角色，水也和清洁联系在一起。这可以从许多基于宗教的元素中看出来，如洗礼池或位于清真寺内外喷泉。在小型足浴、天然泳池或华丽的温泉浴场，水是放松、玩耍和运动的代名词。

意象

水也关乎财富和权力，在很大程度上它甚至可以超过设计的本身，发展成为一个象征，就好像罗马城中华丽的喷泉，凡尔赛宫巨大的水轴，或是在中国新建的雄伟的水坝工程。水可以很宏伟，或是具有象征性，作为设计元素这取决于它如何被应用。它可以在城市中心成为地标，可以在市场中心吸引顾客，也可以在办公机构内发挥着解压和放松的重要作用。

技术难关

几个世纪以来，技术难关的不断解决，像最基础的水的灌溉、治洪等都会加深对于水管理认知的理解。因为当地问题的特殊性，有一些自然和地域特有的技术手段，像修建喷泉、蓄水池、洪水治理这些实例都可以扩充水景设计的技术模式。

水景设计是在一个复杂多样的背景下形成的。它涉及一系列形式、运动、技术，还与现象、神话等相关，因而它能激发无限的幻想和创造力。然而，无论它作为建筑要素的作用有多好，表现出怎样的多样性和魅力，归根结底它只是水。

常见水景分类 **表 1**

类型	自由元素	再生元素
喷水	喷泉	人造喷泉
	间歇喷泉	旱喷
	瀑布	小瀑布
流水	河流	运河
	细流	壕沟
	小河	渠道
静水	湖	盆地
	池塘	污水坑
	水池	水槽
	水坑	鸟浴槽

P11

流水——模型与实例

水是备受青睐的设计元素，它可以通过多种方式进行开发。无数成功的案例证明了水景不仅反映了自然风景的韵致，而且体现了人工建造的高超技艺。

水一直处在不断流动的自然过程中。水景有很多不同的分类方式，设计中最常用的是依据不同的水体流动状态进行分类——喷射、流动、静止和消逝（表1）。

P11

喷水

喷水可以呈现为较自然的方式，如通过泉、间歇泉、雾喷泉或瀑布；也可以呈现为较人工的方式，如壁喷泉或大型激光喷泉（图1～图3）。

喷水量是喷水设计中的一个基本组成部分。其他重要方面还包括：

—喷水压力（例如涓涓流动的小河或充满力量感的强流喷泉）；

—水的体积（一个狭窄的管道或者奔流的瀑布），以及水源的数量与方向（单一的、直线的喷水，或者是一个覆盖更广面积的激光喷泉）；

—出水方式（一个石头之间的小缝隙或掩饰精美的喷泉管），以及水源的周边环境（长满植物的水源或者是充满艺术感的、经过设计的水池）；

—时间间隔设定（恒定的或有节奏变化的设定）。

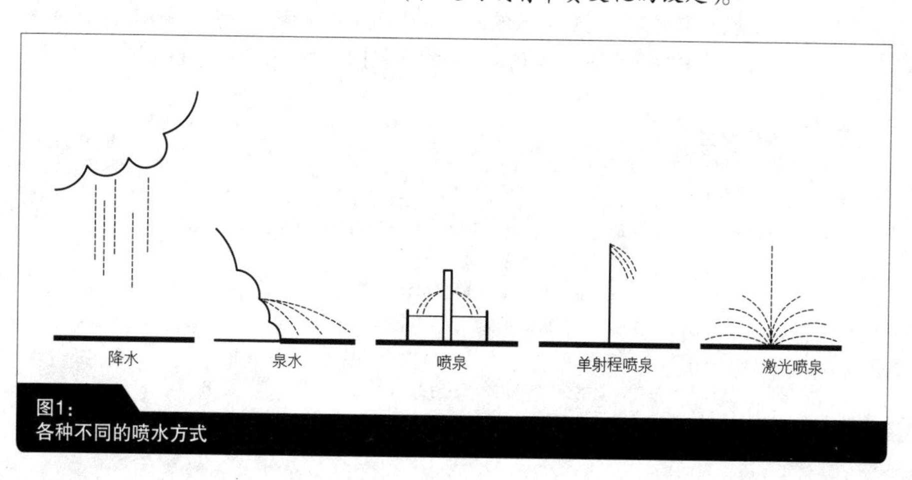

图1：
各种不同的喷水方式

图2：
水从两个石头之间以自然的形式涌出

图3：
设定好恒定时间间隔的旱喷泉

喷水产生了一种稍纵即逝的稚趣图像，也时常伴随雄浑的浮力、轻盈的动作、活泼的声音和实实在在的新鲜感。

包括喷水在内的景观元素突出空间中的一个点（例如一个市场喷泉或者是房子入口处的水池），常常创造了一个独特的空间填充符号。

P13

流水

流水流经狭长的容器或者是存在于一个层叠水池的序列，这些构思通常来源于自然风景的启示，比如绿树成荫的水塘、蜿蜒的小溪或者潺潺流水的石阶。其中人工的元素包括水槽、水渠或瀑布（图 4，图 5）。

示例：

一个极简主义设计中使用喷水的例子，在地面既看不到水池，又看不到喷水口的旱喷泉。Stauffenegger+Stutz 设计事务所利用这种原理在瑞士首都伯尔尼联邦大厦前空旷广场上设计的水雕塑。广场上的每一个喷泉代表瑞士的一个州，轻盈的雕塑、向上流动的动势、定时的舞动序列创造令人印象深刻的画面（图 3）。

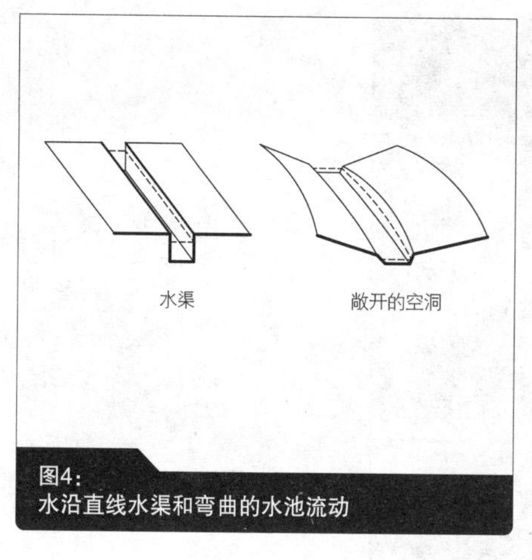

图4：
水沿直线水渠和弯曲的水池流动

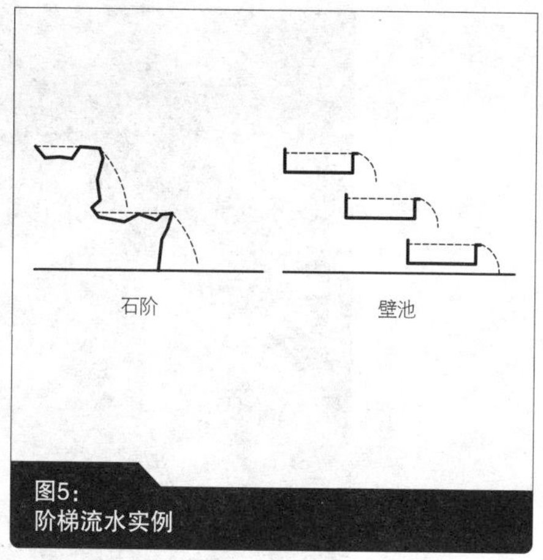

图5：
阶梯流水实例

涉及的设计要素包括：

—流水的量，这意味着设计中要考虑水道宽度或深度及其相关流速；

—水的流向与间隔（例如，通过直的水渠、附加的水池或水坝）；

—遏制和引导水的驳岸（图 6）。

水草丛生的窄堤可以吸引人们将更多的注意力集中在流动的水体之上。宽厚的墙体或粗糙的砾石护岸（例如松散的岩石垒砌，或

图6：
在绿树成荫的有砾石的水池中“天然水”流动

图7：
狭细的水流流经石板之间的缝隙

图8：
逐级叠落的水渠

填充石头的金属筐笼）营造了一个更加清晰的视觉印象。这些堤岸突出了水的力量，并能够创造一种具有粗野风格的独特设计，这种风格随着设计中使用石头尺寸的加大而增强。

在小溪中的流水或瀑布传达一个完整而又富有活力的形象。运动、间歇、水流速的变化总是让人出乎意料，正是这种出乎意料才更易吸引人的眼球。迎面流过来的、潺潺的、飞溅的流水创造了一种明快的、温柔的声音。

然而最重要的是，流水可以让线性的设计要素（例如水渠）得以延续；它连接起空间中的两个点（例如沟渠或者小溪）（图7和图8）和强调地形（以落水和瀑布为例）。

P14

静水

静水的营造需要一个中空的容器或者是一个可以不漏水的水池。这些可以被作为一个开放的生态理念而应用到设计中，例如以浅水池塘、水池或者是湖泊的形式，或者是一种更加几何化的、建筑化的形式（例如：钵、盆或槽）（图9）。

涉及的设计要素包括：

—水岸设计（例如地平面到水体之间过渡部分的设计或者是具有休憩功能的亲水阶梯的设计）；

—水体伴生植被配植（没有植被覆盖的水体、长满断断续续的

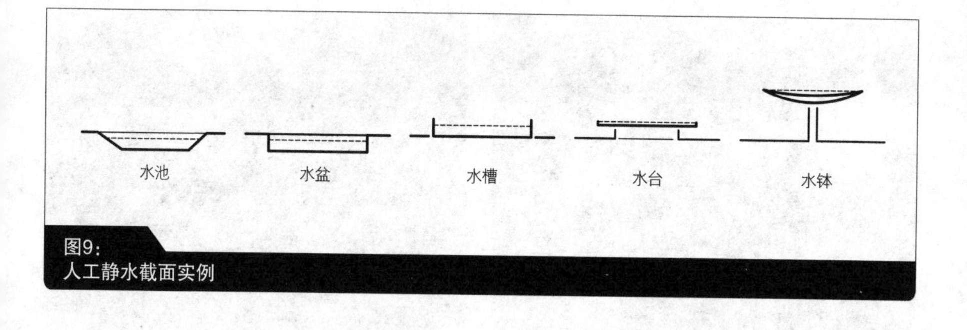

图9：
人工静水截面实例

藤本植被的水体，或者长满茂密的睡莲的水体）；

—灯光的应用及其反光特质的运用。

平静的水面映射出蓝天、捕捉光线而且波光闪烁。这样的水面也可吸收光线，从而给人留下深刻的印象（参见“设计方法”一章中的“感官体验”）。

静水表达的是一种内在平和而有力的沉稳，传达一些静谧的涵义，而且发人深省。

水面的运动是由风或者其他一些外部因素而产生的微妙效果，或者是由于海浪拍击海岸以及其他的构筑物而引起的。

平静的水面能够在具有建筑感的空间中标示出一个中心（例如，在建筑物一层中央能够反射周围景物的倒影池）；静水也可以明确边界（如城堡护城河），或者是作为一个微妙的定位系统（湖泊作为一个方向点或者作为一个公园内沿着游览者的足迹的转折点）（参见“设计方法”一章中的“功能要求”）。采用静水作为水景设计对象往往比之前提到的其他水景观（如喷水、流水）具有更大的空间需求。

示例：

在德国巴塞多景观公园中，彼得·约瑟夫·莱内（Peter Joseph Lenné）正在研究一个处于视轴中央的、具有沿着水体边岸略有加宽的小水体。绿色的区域和树丛间断性地遮挡整体画面；开阔的水域体量似乎变得更加巨大，而且游览者愿意看到的是不断变化的形式和深度（图 10）。

示例：

在法国沃乐子爵（Vaux-le-Vicomte）公园中，勒诺特（Le Notre）在具有明确边界的树篱包围的场地中央设置了一个自然形态的静水水池。树的光影、水面光斑的变化有效地消解了生硬的、明确的水岸边界，由此形成了人工布局和自然环境之间的对比（图 11）。

图10：
一个设有多个入水口的狭长水体的景观公园

图11：
一个由树篱和树林围合的几何形水池

P16

逐渐消逝的水

水的消失是水循环的最后一个部分。水能够通过排水系统排掉，例如，水通过阳光和风从物体表面蒸发，或者渗入多孔的地面，或者蒸发变成雾。

设计的方向包括：

—不同的流水消失可能性的考虑（装饰格栅或水槽放大声音）（参见“技术参数”一章中的“进水和泄水”）；

—时间和速度之间的相互作用（在沥青路面上形成的暂时性积水池，或者缓慢地从砂层渗入地下）；

—在偶然、短暂瞬间的水景营造（雾喷或水分的蒸发）（参见“技术参数”一章中的“水体流动”）。

包括日常蒸发水在内的概念设计具有一种独特的魅力。这些设计在薄雾之中笼罩着充满异域风情的植物、错综的道路、洁白的临时展示空间，或营造漂浮在湖泊或池塘上可供游览者体验穿越的云雾，以这种方式设计的作品是独特而迷人的，因为它们赋予了人们熟悉现象以全新的外在感受。

将水分蒸发作为设计元素的作品并不是很常见。它们通常是非常脆弱的，形式和边界都十分的庞大，而且它们的特性也很难预测，往往需要更多的经验积累，这就是为什么它们几乎在公共空间中不

图12：
刻意不设中心的倒影池

常见。但也正是由于这一点，再加上非常容易引起人伤感的瞬间消逝，才使它拥有一种独特的诱惑，值得去探索和设计。

P18

其他方面

水元素不会受到以上四个方面局限，可以组合开发（例如，从地面喷出的泉水流入溪流之中），或者利用其中特定的环节。

水的缺席

水的“缺席”就是一个例子。这可以被许多案例所证明，例如，结构和材料的使用与水的存在直接相关（例如，石头的安放或者是干涸的河床）（图 13）。这也能够通过自然的元素或抽象的手段被象征性地暗示（以铺满砾石的倾斜的面或者多草的区域为例）（图 14）（例如若隐若现的太阳能电池板或者黑色沥青的表面）。

示例：

在瑞典温特图尔的 Steinberger Gasse，艺术家托马斯·斯奈德霍普（Thomas Schneider-Hoppe）和唐纳德·贾德（Donald Judd）正致力于一系列简单圆形喷泉的研究。他们运用一种逐级下降的和非常精准的方式，采用流水和消失的水进行创作，而且用没有中心的倒影池引起观看者的好奇心（图 12）。

图13：
在干涸的河床中一个明确的水的“缺席”

图14：
裸露的白砂象征水的在场

水的缺席可以使人们头脑中水的图像更加清晰、诱人，也时常发人深思、让人心灵宁静，而且具有引人冥想的特性。

水景总有一段时间会被停用（由于天气的原因或者是被限制时间来运用），或者如果它被暂停使用时，可能由于安全问题或者是维护的限制，水的不在场则是替代一种水体景观来使用的。

冬季影响

即使在结霜期间，水的独特效果仍然能够被纳入到设计中。白色、柔和、釉面的霜能包围灌木丛或者树木；雪可以消隐部分设计元素，抽象视觉本身。在瀑布或者小瀑布上，还可以转变为独特的形式与图像。一个光滑的、平整的冰面，例如那些倒影池或者池塘能够被用于冬季运动，如滑冰或是打冰球。

示例：

日本园林中非常美丽的枯山水，均衡布置的石头（象征着岛屿和水岸），剪形树木（象征着森林），波纹状的细沙（象征着水体表面和波浪），隐喻着自然的意趣，这种缩千里于方寸的艺术，在小尺度空间中表现得淋漓尽致（图 14）。

图15：
在后工业景观中心的巨大倒影池在冬季时成为临时滑冰场

即使这些仅仅是辅助的方面和使用的可能性，例如包括漫长的冬天，这些都要取决于当地的气候特征。冰封水面形成令人印象深刻的景象，也应该从技术的角度来考虑（参见“技术参数”一章中的“冬季防护”）。

示例：

在德国埃森 Zeche Zollverein Landscape 公园，一个大规模的、旧的工业结构在 12 米宽的带状倒影池中被凸现了出来。在冬季的几个月内，在相邻的冷却装置的辅助下，它成了一个滑冰场（图 15）。

图16：
水面与池沿持平的喷泉

P22

设计方法

水景设计是一种极具个性化的实践，这种实践受很多因素影响。这些因素中涉及对水景元素具体的处理手法、场地的实际情况、作为建筑元素在空间中发挥的作用、需要完成的功能作用、各种各样的感官因素以及水所能表达出的象征性力量。

P22

水景创作

创作一个水景，要处理好水景独特的活力特性、设计元素与观赏者之间的高度关系（或视觉关系）、历时性的体验，以及它所包含边界（比如河岸或水池边缘）的形式处理等这些因素。

内在活力

水景运动的方式和方向——是设计中最为本质的地方。根据水景的具体类型和特定的设计场地，设计者需要考虑是运用静水、流水、瀑布还是喷泉，并控制适当的水景活力、水量、空间分布以及流速。

视线关系

水景元素的布置与视线高度之间的关系——决定了这个水景能给人怎样的体验，也因此决定了所选设计方法的整体效果。一个低的感知角度能够让人对水景进行较好的概览；一个稍高的感知角度，如膝盖的高度，能给人更多形态上的体验；再高些的视线水平能给

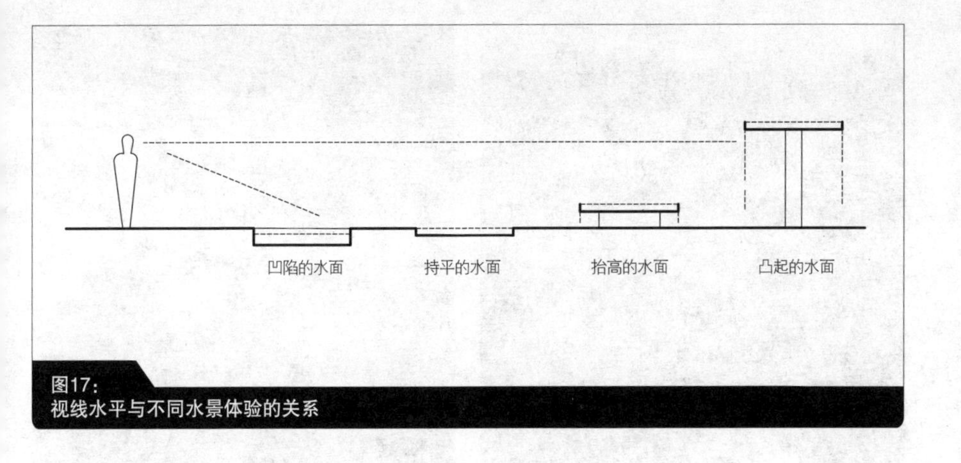

图17：
视线水平与不同水景体验的关系

人以很强的距离感（图 17）。

用低洼水域造景，能够给人以很好的全景视觉效果。在蜿蜒的水岸会遮挡住大部分实际水面的情况下，这种手法同样可用于处理水位有显著波动的水体（比如水库）。下凹的水体具有清晰的边界。对于规模较小的水体而言，这些边界在比例上会显得过大，因而会限制水景的体验效果。此时，为避免这样的问题，要将水面向上提升，创造出微妙的变化。这就需要设计者对岸线进行精确的设计与建造，尽量减少岸线的水位波动。

要加高突出的水景元素（比如说水池和水钵）可以加高到 0.4 ~ 1.2m，这样的高度较为平易近人；或者加高到高于视平线的高度（大约高于 1.6m）。由于透视收缩变形，一个升高的水体看起来就会显得比位置较低的水体小。加高突出的元素让边缘显露出来，在不同高度的情况下，有时甚至可以显露出底面的支撑结构，因此，这就需要针对形态方面、材料、观赏性、溢水量以及水景类型作额外的设计考虑。

运用喷泉水景时，比如单射程喷泉，可以通过距离特效来强调特定的方面，这能给人们带来更好的视觉体验。

时间体验

静态水可以按照一个时序结构来设计安排视觉关系层次，比如说高低错落的静水池。而动态水景能为其他视觉感知高度和瞬时性因素的产生提供契机。水景的空间分布、动态强度、水景的时序控制（图 18）能够形成交替的、瞬时的景象（参见“技术参数”一章中的“水体流动”）。

图18：
间隔性临时“降雨”装置

图19：
具有书法风格的带有角度的水体干预构造

水池

选择自然水塘还是人工水池，与设计的焦点直接相关。水景的设计，可能要涉及光、动态以及深度的处理，比如设计映像池中独立的单孔喷泉。同时，水景还能在周围环境中得到烘托，通过与建筑的对比与对话更添魅力。

除了喷雾喷泉外，水只能用水池元素汇集在一起运用（比如水槽、蓄水池、不显眼的狭窄流水道），而且水的形态对于设计对象的整体特色而言有着巨大的影响（参见“技术参数”一章中的“水岸设计”）。在水体的结构和设计元素、观赏性喷泉水池和造型装饰物中都可以看到这方面的例子（图 19、图 20）。

P25

场地独特性

水景是不能被单独创作出来的。它是建筑群中的一部分，是存在于一个复杂的、具有空间性和主题性的关系当中的。

场地环境对于一个设计元素的效果起着至关重要的决定性作用。一个同样的喷水池，放在不同的场地，呈现出的效果完全不同——放置在鹅卵石广场的中央，在四面环墙的院子里，在杂草丛生的灌木园，又或者是在开放景观中。场地与它特有的品质，场地中的支配性关系，历史脉络以及空间形式对于一个有逻辑性的创意想法或一个元素的设计质量是非常重要的。

图20：
雕塑组合喷泉

图21：
在新景观公园中心处的旧过滤池

独特之处

在这些假定的限制当中存在着巨大的潜在设计空间。如果有充足的资金，几乎所有东西在技术和创意上都是存在可能的—— 一种随心所欲、可不断更新设计作品的自由。重要的是要找到一种独特的思想。这种独特思想应当对设计方面的问题具有一定意义，而不仅仅是与预算有关；要坚定地投入到场地与场地潜力的体会中，运用这种感受创作出个性化的景观，拿出可持续性的技术解决方案。

先决条件

城市结构可以提供重要的参照点。场地中的重要路径或者周围建筑都可能成为场地内的主导要素。早在概念设计阶段，场地内先决要素，比如雨水管理或者初步设计计划，就已经对作品的形式产生了影响。

历史遗留

地形结构与以往的使用情况以及地形独特的历史痕迹都有关系。因此地形结构就可以成为设计的其他参考点。景观结构提供了独特的形式特点和其额外的历史叙述性。古老的喷泉或水道遗迹可以被重新利用，杂草蔓生的溪流清理一下，像工业文明这样的遗迹也可以被纳入新环境中的（图 21）。

空间形式

场地的空间情况，其建筑结构和现存植被是整体设计的参照。

举例来说，在巴特恩豪森，有感染力的水景元素可以在一片空地上打造一个醒目的中心点（图 22）。水景元素的存在与美丽可以发出共鸣，也可以渲染氛围（图 23），或者表现出与周围建筑的会话时

图22：
大型间歇性水景喷射池

图23：
反射池凸显了场地的沉思性与冥想性

作出自己独特的回应（图 24）。

形式上的呼应究竟是受自然美景的激发，还是一个更加结构化和建筑化的交流，是由环境因素决定的。面对一个规模庞大的场地时，应当以景观效果为重。简洁的空间，比如一个院子或者其他封闭空间的情况，是更易于用建筑化的形式语言和抽象化的方法去处理的。

P28

功能要求

在涉及水景的设计中，可以通过巧妙的组合和布置来达到功能要求和特殊目的。除了给水、排水的诸多方面的问题，还要考虑娱乐活动和体育运动方面的问题。

饮用水

饮水器的基本功能是供给饮用水。随着将独立公寓包括在内的公共饮用水供给系统的发展，公共饮水器为城市供水的基本功能已

示例：

在 Latz and Partner 设计的杜伊斯堡北景观公园，以前的废弃钢铁厂的过滤池被清洗干净并用水和繁茂的竹丛将其填满。现在，这里形成了地面上一个独具特色的景观焦点（图 21）。欧洲最古老的人工潜水中心新建一个人工水底世界，就位于煤气厂附近。

示例：

在柏林弗里德里希斯海因区（Berlin's Friedrichshain district）一个废弃工业中心的院子中，古斯塔夫·兰格（Gustav Lange）设置了一块带有细窄流水道的石灰岩。通过与四周砖墙的对比，这个石块的形式和比例营造了一种简明、独特并又强而有力的感觉。静水与生长缓慢的苔藓和蕨类植物更为这里增添了浓厚诗意，同时又能与周围环境相映成趣，让景色生动迷人（图 24）。

图24：
在建筑环境中带有流水道的大型石灰石块

经退化。今天的饮水器中已经注满了工业用水，只作纯粹的装饰之用。饮水器在温暖的天气中会更受欢迎。温暖的天气中饮水器是与饮用水供应系统相连接的，在城市中心或娱乐中心和体育设施附近都十分常见（图25）。

雨水管理

雨水落在封闭表面上，比如屋顶或街道，那么雨水就需要被收集起来妥善处理，以便长久地保护建筑物。比起把雨水引入一般排污系统，从经济和环保角度来讲更明智的做法是让雨水留在它着落的地面，然后通过蒸发和渗透将它引回水的自然循环系统。雨水排水管道以及汇水渠（图26）、植草沟（图27）、水塘、地下集水池都可以起到这个作用。设计者在户外设施技术结构的集成和设计方面有很多自主权，可持续水资源管理对于新开发区域，复杂而广泛的雨水管理可以在建设之初就可以实现。

提示：

想了解基本信息和其他设计理念，可以参考本套丛书中的贝尔特·比勒费尔德（Bert Bielefeld）与塞巴斯蒂安·埃尔库里（Sebastian EL Khouli）编著、张路峰翻译的《设计概念》一书（中国建筑工业出版社，北京，2011年）（征订号：20273），以及Hans Loidl和Stefan Bernhard写的《Opening Spaces》一书（Birkhäuser Verlag，Basel，2003年）。

图25：
设置在下沉式水池前的前卫柱状造型饮水器

图26：
带有防溅装置的雨水管

排水　收集的雨水和喷泉水可以用来灌溉花园和种植园。研究表明，太阳暴晒几日，被阳光晒热的水会更有利于植物良好发育（图 28）。用于收集水并供水的水池、植草沟可以被很好地整合到设计中，利用集合优势，发挥雨水的多样用途。

娱乐和运动　水对于娱乐和运动来说是非常重要的，尤其是在公共公园中。在适当的条件下，现有水资源可以被拓展与整合到户外场地中。这里有一个很好的例子，就是慕尼黑的英国花园（English Garden）。在规划设计阶段，斯开尔（Friedrich Ludwig von Sckell）创造性地用一

图27：
通过斜插入钢片形成的草坪中的雨水滞留池

图28：
在植被密集的场地前的集水池

图29：
在慕尼黑Eisbach河的人工冲浪

图30：
游泳池沐浴区的人工水池

图31：
具有多种深度的游泳池

图32：
设有手动水泵、戏水场地和玩具挖掘机的游戏场地

种类似景观设计的处理手法，加入人工波浪，将现有的 Eisbach 河纳入到设计方案中。今天，其设计形式受到广泛认可，英国花园也成为一个著名的游泳和冲浪胜地（图 29）。

即使是一个自然式结构、经过景观规划与水处理的人工游泳池能够轻易融入绿色区域。不过，这种做法只适合较低或一般的使用要求（图 30）。如果使用要求较高，就需要一个拥有电子水处理系统的加固水池了（图 31）。

水上娱乐区拥有多种用水形式。用水泵、阿基米德水螺杆以及水车供水进行发挥创造；用流水道、水箱、风车叶轮来优化水的分

图33：
附带码头和引导游客的桥梁的护城河

图34：
作空间分隔和声音阻隔之用的水景墙

布并加以引导；同时要创作出一个将水、砂子和泥土结合到一起的设计作品（图 32）。

视线的封闭和引导

水也可以设计成一种障碍物。又宽又深的壕沟，比如城堡周围的壕沟，是可以替代墙或者栅栏的。这种壕沟可以起到阻隔作用，却不会阻断地面上不同区域的视线连通性。位于水体之上的桥梁和码头能够连通道路，与作为障碍物的水体一起，共同形成一个规划导向系统（图 33）。在历史公园或现代游乐园中的规划，像系统的中心和焦点，常常受到不可跨越大型水体的限制，比如湖滨和峡湾。游人会沿着既定的路线被引导至重要的场地或景点。

相比之下，高大的喷泉或者水幕拥有遮挡视线的效果，它们形成一个保护层可以隐藏起不需要看到的功能和元素，它们所产生的声音还可以掩蔽来自邻近街道上的噪声（图 34）。

P31

象征主义

水有丰富的象征意义并且具有较深的宗教渊源。这种现象已经产生了几个世纪，时至今日也仍然适用，即使是在无意识层面也存在这种现象。

象征物

根据周围不同环境，水景不同的形式和运动状态可以象征宁静安详、心旷神怡、活力生机或者财富价值。它是生命和瞬间的跨文化的象征。水不仅是生存的基本保障，同时也是人类智慧和精神的象征。月亮、水和女性在象征意义上紧密相关。

宗教中的水

世界三大宗教都是发源于干旱气候地区，因此，不用说，大自然的宗教意义自然而然地从一开始就与水元素密切相关。水至纯至净的品质常常得到赞誉：在伊斯兰教的净身仪式或进入清真寺前的洗礼仪式中，又或者是在印度教所信仰的在恒河沐浴中。几乎每个犹太人社区都有一个干净流水的浸礼池或礼仪沐浴池。但是，只有那些将自己整个身体浸没在水中的人才算在沐浴仪式中得到了净化。

长久以来，基督教通过将身体完全浸泡在水中或将水泼到身体上来让身体得到彻底的洗礼，以此方式进行洗礼仪式。在西方社会里，这种习俗已经逐渐简化为仅仅将水滴淋到受洗礼人的额头上。洗礼象征着对基督耶稣的献祭以及被教堂所接纳和认可。洗礼也同样象征着耶稣的死亡与复活。在天主教和东正教中，圣水通常有着非比寻常的意义。

P32

感官体验

很多元素具有超越单纯的功能层次—— 一个独特的特性，设计方案必须对这种特性作出特殊考虑。举例来说，植物就可以从第四维度——时间的角度来诠释设计元素。随着四季变换，植物的成长以及在形态、色彩方面的变化会年复一年地改变景观设计作品的面貌和气氛。水的特性会以千变万化的方式存在，你可以通过感官来体验这些特性。

如果能将感官体验融入设计理念中，拥有水景的设计会变得更富表现力。很难说出哪种感觉更多或哪种感觉受到水景的刺激最多。但是归根到底，一种和谐、均衡的制衡互动才是一个作品成功的关键。

视觉与色彩

纯净的水是清澈而透明的。这样的水能够吸收光线。水滴、倒映水池以及水浪都可成为棱镜，在这些棱镜中，光被折射并衍化成许许多多缤纷闪烁的光彩。沉积物、溶剂或乳化剂能够让水变得色彩缤纷。

水中的气泡会让水看起来不再那么清澈透明，变得泛白而浑浊。当水体平静，水中的空气就会消散，这种特殊的效果也会随之消失。就是这些水中的气泡塑造出了白色的浪峰以及泡沫型喷水枪或“schaumsprudlers”泡沫型喷嘴制造的柔软的白色水雕塑（参见“技术参数”一章中的“水体流动”）。

比如说，那些水体流经土壤和石块时夹带的物质，也可以赋予水一些颜色。常常发现荒原地带的排水口周围会染上些棕色，这种

图35：
东方园林中的喷泉与水道

图36：
修道院附近一个造型简单的喷泉

着色来自那里普遍存在的腐殖酸淋溶剂。来自溶剂的颜色耐久，且能够在很长一段时间里保持稳定。但是这样的颜色暗淡且难于控制。

汹涌湍急的流水从河床上撕扯下砂子和石块并且将这些砂石卷推向前。这些沉积物为水体增添了色彩，但没有我们平日见到的那些鲜活、清新的山溪那样清澈透明。水波平静后，那些沉积物沉淀下来，它赋予水的色彩也随之稳定下来。

然而，水体的颜色常常来自于水体表面对周围环境的映射或来自于水下的物体。我们对这些色彩的感知形式取决于视线角度，以

示例：

水象征着生命、死亡以及人类的复活。举例来说，样式沉静从容的喷泉常被放置在基督教墓园或附近的修道院里，如安放其中的细缓的喷射水流、石雕喷泉或小型水迷宫等水景形式（图 36）。东方园林中，水代表着天堂中的四条河流（图 35）。

图37：
附有扶手的池底覆盖物的“克奈浦”水疗池

及光线亮度不同时段在空气和水表之间的折射角。水越深，水的颜色也就越深——在河岸和河床上形成的自然沉积物会加强这种效果。暗沉得几乎难以辨认的池塘底部能够增强水面的反射效果。像游泳池那样浅色的底会让水看起来更清澈、明亮和透明，而在池底的那些装饰也能因此而被容易看见（图 38）。

味觉

纯净的水本身是没有味道的。周围的土壤或岩石的淋溶和溶解会给水体带来一些气味。比如泉水就可以被开发使用，接入饮水器。

嗅觉

“水之香气”和水的味道产生原理相同，都是由可溶性添加物产生的。为了让水的气味能够被感知到，具有芳香气味的水需要以蒸汽或水雾的形态释放到空气中。在自然式瀑布周围那些清新又混有些微矿物质的气味正是以这种方式产生的。

示例：

“克奈浦”水疗池是用齐膝深的冷水沐浴的水池。水流的上下波动和冷热水交替以其保健作用而闻名，辅以像砾石这种能够刺激触觉的池底覆盖物，保健效果更佳（图 37）。

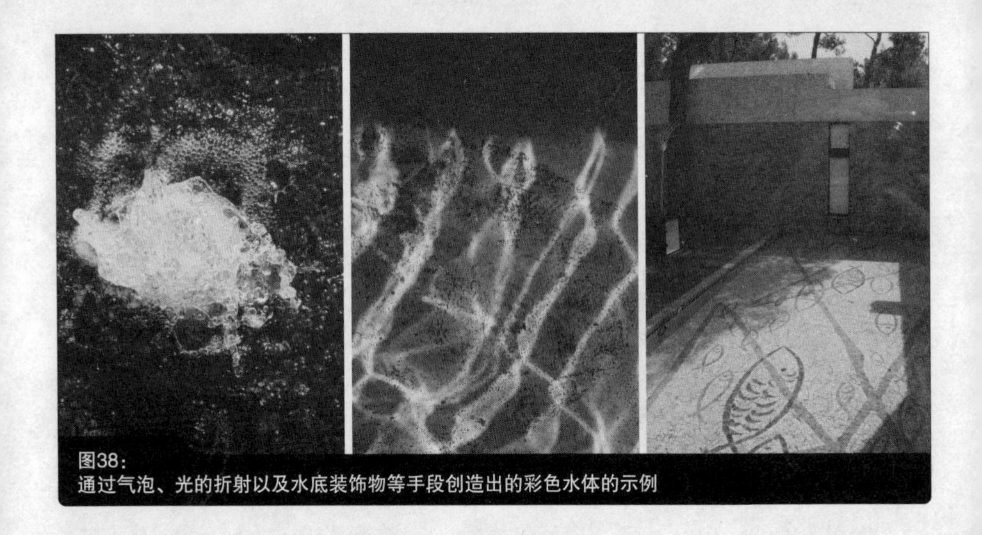

图38：
通过气泡、光的折射以及水底装饰物等手段创造出的彩色水体的示例

触觉

人们可以通过潜水的形式来直接与水进行物理接触，或通过接触蒸汽或水雾进行间接物理接触。温度和空气湿度可以用来比较设计效果。比如，雾化水在干燥炎热的环境里是清爽宜人的，但是在一个凉爽潮湿的环境中就显得让人不适，寒冷而不为人喜爱。

听觉

水的声响高低起伏，富有节奏又不断变化的音调，展现出很多音乐的品质：山间河流的呼啸，喷泉洪亮有力的流水声，泡沫型喷嘴的振动，或一滴水的坠落。音量与音质取决于水的量，声音的速度、产生共鸣的位置以及水从何处坠落。因此，水流的速度，水道的类型，水源的状态，冲击面的面积、高度和深度，水泵循环的频率都会对水景特有的声音产生影响。

提示：

由于一些因素的影响，水景声效就像水体降落一样难以控制，并且需要在施工阶段进行测试和调整。这种测试和调整，可以通过对法兰（如调节水体流入的不同角度）、共振体（如用石头来代替木头）或者水体流速的调整来完成。

图39：
结合动态喷泉、人工瀑布、静态水池等水景类型的“水风琴”音乐动态喷泉

图40：
独具特色的水景设施

示例：

意大利蒂沃利的埃斯特庄园中的许多动态喷泉，人工瀑布以及静态水池等的形态，都是从乐器形象中提取的视觉元素，同时通过改变水压、喷射流的力度以及承接水流冲击的不同表面来发出不同的声音（图 39）。

P37

技术参数

基于技术可用性和资金可行性，设计师在构思出一个基于现有场地和个人灵感的水景设施的概念后，必须将其落实到可持续设计项目中。

场地问题及解决方案是多样而个性的，需要设计师具有足够的想象力，并且具有将构思细化到最终细节的能力。水的可利用性、资源的问题、水池的类型、水的流向、运动及整体背景透视关系的设计都是要考虑的重要方面。

P37

水源

水作为布景

在设计水景设施之前，要确立的第一件事是：在水景设施运作期间，是否有足够的可用水源。水可以从喷泉、湖泊、河流、雨水、地下水或当地管道中获取，并在需要时保证供给。

有足够的蓄水能力，并且有溢流点的天然喷泉是理想而经济的水源。水的品质是由水流经的土壤所决定的，在喷泉出水口处建造一个遮盖物以保护其免受污染，也可以用联结多个喷泉的方法来保证足够的蓄水量，这样有助于调整单个喷泉由于一整年的涨落形成的不同供水能力。

水源也可从地上资源中获得，如带有管道、水泵或水动力装置的溪流或湖泊。有自己特定的水坝和尾水渠的工厂也可使用相似的方法。这保证了水供给的持续性，但水质往往需要更广泛地处理。

水的凝结态，如雨或雪，可以常年从屋顶、城镇广场或其他封闭表面中进行收集。收集的水量、分布的方式和时间由当地气候决定，并且有很大的变化。这意味着水景设施的水供给将出现短缺现象,尤其是在夏天或者干旱的延长期。在被收集和季节性的使用之前，收集到的降雨均储存在地上水池或大小合适的地下蓄水池中。雨水的水质通常很高，但会被临近封闭表面的溢流杂质所污染。

地下水可以在透水岩层和地下土壤层中发现，并很容易从可渗透的地表中泵出，如砾石和沙土，这样就保证了水源的持续供给。在砾石开采过程中，由于大范围的挖掘而显露出的地下蓄水层，可利用为永久性的、高品质水体的湖泊。

有一种简单获取水的方法，即使用现存的饮用水系统。水的供给通常是持续的并且水质很高，不需要额外的处理。但是所需的花费也会很高。

水景设施可以利用单一的水源，也可以采用不同水源的结合，其决定因素包括当地水的可利用性、现存蓄水池内水的可利用率、产生的费用和预期达到的品质。

结构

水景设施（图 41）所需的水是通过注水和溢流来提供的，偶尔在连接处带有一个相接的蓄水池（参见“技术参数”一章中的“水的注入和溢流”）。水可以单独存在（如自给的喷泉）或者与配置的蓄水池一起存在（如人造喷泉）。水池是凹形的，有的可见（水池、盆地或山谷），有的不可见（在格栅下面或在泵房的下表面），这取决于设计（参见“技术参数”一章中“池底垫层和水池”）。并且可以另外规划内部液压循环系统以运转水（如净化循环水）（参见“技术参数”一章中“水体流动”）。

水量

水景设施所需水量是由主体设施容量（喷泉水池或池塘）、隐蔽设施容量（连接管道或缓冲池），以及暂时用于倾泻或喷发水量共同决定。

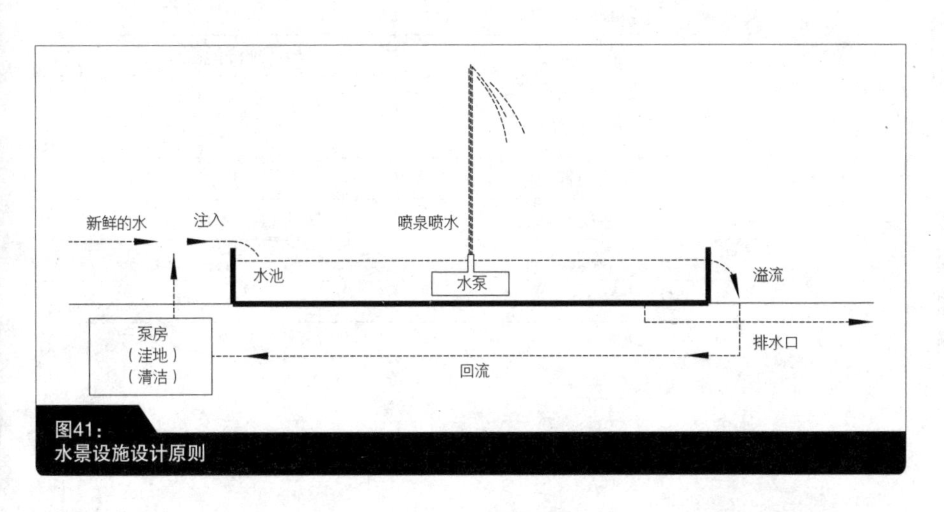

图41：
水景设施设计原则

提示：

在大多数国家，对现存水做任何干预或改变都要经过严格的法律。这也适用于地下水的使用。流程的申请和许可证的颁发所需要的时间通常很漫长，在方案开发期间就需要考虑到这些问题。

提醒：

最好在包含与水有直接身体接触的设计中使用可饮用的水。指定的饮用水喷泉和水景设施，如与水直接接触的戏水区，由于卫生的原因不得不使用饮用水。

暂时性的注水是为了完全填满水景设施，持久性的注水是为了补偿水的流失。常规使用将造成水的流失，例如，水的涌出、喷泉喷发、补给蒸发、为维持水质而替换的一部分水。

P39

池底垫层和水池

水景设计无论在概念上还是在建造中都容不得差池。水能找到最微小的渗水点，通过渗水点止步于下一个不透水区域，这一过程令人难以置信的精确。

合适的水池或足够防渗的地基土层在工程项目中较少遇到。因此，安装持久耐用的水装置时，需要考虑的因素包括：选择耐用的池底垫层；总体设计时期和过渡期要有精确而详细的规划；各组件之间应有稳定的衔接。

不同的池底垫层　　表 2

开放性池底垫层 自然式景观	连续的池底垫层 人工式景观
黏土 / 淤泥	现浇混凝土
膨土岩	混凝土构件
柔性膜	塑料
焦油沥青面板	钢
沥青胶粘剂	木材
喷浆混凝土	砖石
塑料	自然石材

提示：

暴露于太阳和大风下引起水的蒸发的损失，即使在温带气候区，一个开放的水体每天能达到 1 厘米的水位下降。要解决这个问题，设计者要确保水体在一些阴凉的地方并尽量避免被风吹到。

由于池底垫层材料不同，其技术参数各不相同，更由于设计者构造理念的不同，因此导致了每一种池底垫层各有特点（见表 2）。

决定池底垫层类型的因素有：外观、环境、形状、尺寸、水运动所需的能量、水体流动的方式、土壤的性质等。池底垫层的顶部边缘必须始终连续，远高于预期的最高水位（参见图 42 和“技术参数”一章中的“水岸设计”）。

地基土层需要充分的夯实，以承载所填充水体的重量。任何后期改造装置都可能导致池底垫层或堤岸设计的损坏。因此要按规则使用以下方法：

黏土

黏土池底垫层是最古老的方法之一。将 30 厘米的黏土或淤泥加灌到平铺石头的基层之上，然后压实，并覆盖砾石砂保护层。黏土层透水率较低 $K \leqslant 10^{-9}$，当坡度达到 1/3 时可以使用此方法。

池底垫层均匀分布着不承重的大块干燥土坯。对于较小区域的水景来说，土坯砖和预制黏土池底垫层是不错的选择。可以采用多层叠加，然后夯实的方式建造。在施工期间，黏土必须保持适当的湿度以确保安全密封。

黏土和淤泥的自然属性保证了部分潜在使用者在情感上更容易接受。即使装置被拆除，这种材料也很容易出售。但是，如果停止水供应，材料就会完全干燥，甚至会出现深深的裂缝。这些可能会对池底垫层造成永久性的损害。在较小的水景中，黏土或淤泥制成的池底垫层对自身膨胀所产生的变形，有自我修复功能。另一方面，在较大规模的水景中，黏土垫层汇集或贯穿起来是比较困难的，这些通常需要额外的、大范围的柔性膜进行连接。

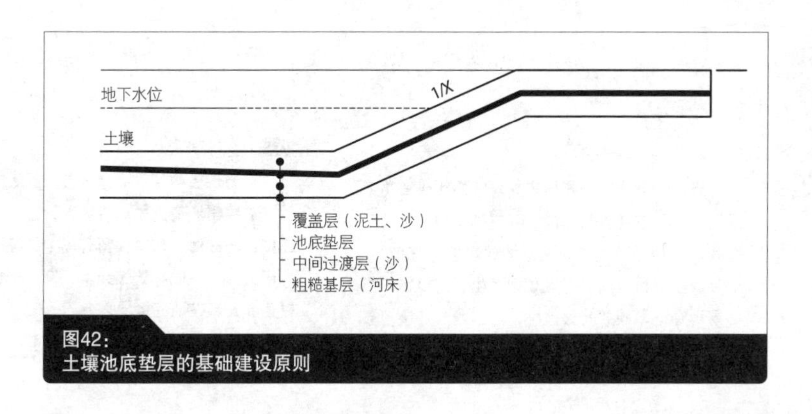

图42：
土壤池底垫层的基础建设原则

这种池底垫层的方法适合接近天然水体的水景、长期满水的水景，或者没有被过量使用的水景，如花园池塘。

膨润土

黏土池底垫层的一个特殊形式，是膨润土。膨润土是一种由黏土矿物制成的高强度吸收的石头，首先被研磨成类似土壤颗粒的粉末状，然后覆盖上砾石或沙子的保护层进行夯实。有些建造商采用一种替代的方法，这种方法是利用松散的混合物取代预制垫，将混合物铺置到土壤中即可。

膨润土可增加土壤实际的抗渗水能力，成为基层上额外的池底垫层，从而省去了土壤替换所需的昂贵费用。如果场地的土壤已经具有高度的不透水性，这种方法很好用，其他应用程序的选择和限制是根据黏土性质而定的。

柔性薄膜

该方法在家庭花园最常见。1.5 ~ 2.5 毫米厚的塑料薄膜结合在一起，放置在预制的平缓的细粒度基层上，位于找平沙层上，坡度最大可达 1/3。陡峭的斜坡会导致覆盖层的侵蚀，由于其压力和紫外线的敏感性会使薄膜逐渐显露出来。

自然的边界可以通过轻轻翘曲的膜塑造成，也可以用黏土池底垫层的方式。用边缘式终端封条可以做固定设备与插入设备连接之间的防水，而后续的修补只能是在一定程度上缓解问题。

柔性薄膜的优点包括：适应性强、安装方便快捷、可渗水地基的密封度以及相对较好的性价比。

此方法很少用在较大的技术构造中，多数用在小场地内的池塘，例如雨水蓄水池。

焦油沥青板

焦油沥青板作为池底垫层与柔性隔膜相似。焦油沥青板两侧涂有沥青，由片材组成，原材料容易获得，使用简单。然而，它们不

提示：

当选择薄膜时，要检查其对紫外线或植物根系的阻力。在覆盖材料前，膜应被平铺，平滑无杂物，因为它可能被撕裂。如果场地内有植物，如甘蔗丛或竹子这类有发达根系生长的植物，很容易地穿透薄膜并将之破坏，可以通过在该植被地区内添加防根穿的塑料池底垫层来防护。

提醒：

在沥青板中输导或泄水最好的做法是，将法兰直接置于沥青之中。将这种构造与构筑性结构连接在一起时，需要保留有一定的缝隙，再用沥青密封。小型的构筑性元素，例如楼梯或踏步可以直接安装在池底垫层之上。

图43：
无缝铺装过渡到路面

图44：
沥青、混凝土和钢材的混合构造方法

耐紫外线或植物根系的生长性破坏，这意味着这种经济性的池底垫层只能被用于临时设施。

沥青胶粘剂

沥青胶粘剂是一种高密度的沥青物质，加热后成为液体，利于浇灌。两个 1 厘米厚的覆盖层铺设于完结的基础地面，带有防冻层、矿物基质和一层沥青胶粘剂。该方法允许≥ 1/2.5 的坡率。这是一种持久和稳定的池底垫层方法，可以在后期进行添加和重新密封，所有程序必须使用正确的技术来完成。这是一个复杂的过程，对于小型装置来说相对昂贵。

石材填充的覆盖层可以很恰当地替换这种方法。生产和安装时更经济，但（同等份的）剂量跟沥青胶粘剂比，不会产生相同密度的孔隙，因此需要沿着四周安装一个技术上更复杂的装置。

石材填充的沥青池底垫层方法建议用在规模较大、地势不平坦、多孔但稳定的基层上。它可以平滑过渡到周围园路的路面，因此可以被用于建造浅的映景明湖（图 43）。

现浇混凝土

现浇法要求水景设备必须建在稳定的地基上，采用高品质的防水水泥，最好加固。用符合规范的模板、起坡角度和准确的形式就可现场铸造（图 40）。由于建筑工地各不相同的工作条件，模板的具体构造物和水泥的收缩强度，直接影响这些构件的大量应用。面层可采用光滑的清水混凝土、油漆、地板或瓷砖。

混凝土无毒，所以对于动植物来说是安全的。初始阶段地表径流会改变水的 pH 值，但换水就可以解决这个问题。

图45：
利用原地混凝土设施制成的喷泉

图46：
精确建造的混凝土构件

这种方法已经被大量应用于大型建筑中，也在游泳池、水渠或水槽这类几何形的工程构筑之中大量使用（图45）。

混凝土构件

混凝土构件在工厂中制造，搬运到建设场地，然后安装到建好的地基上。大型混凝土构件由很多部件组成，尤其是在建设大型基地板的时候，可以用现浇法在场地中制作。

组装构件的结合点必须进行处理，并且要格外小心地将其闭合。除了要确保混凝土的高品质，单独的构筑物也要保证建筑构件的精确性，因为它将影响尺寸的容差（图46）。运输方法和运输容量是唯一能限制构件尺寸和形状的因素。

提示：

大于5米的混凝土分割需要扩大接合点，可以通过插入不透水层来实现。由于这些接合点将影响水景设施的外形，因此接合点的技术设计非常重要。例如，是否把接合点置于规则间隔中等。即使通过针刺或淘析来处理表面，加固物仍然需要进行保护。对于混凝土水利工程建设来说，建议使用大型的混凝土遮盖物以保护钢制加固物免遭腐蚀，这意味着许多建筑物都将比设计预期的规模要大。

预制混凝土构件的元件称为铸石，其可视表面是石质的断面，或者设计为针刺的、颗粒状的、喷砂式的、蚀刻的、水爆破式的或砂质的断面。混凝土构件通常用作巨石元件或一种特定需要的表面构造，量大时需在场地内生产制造，但要远离建设区。这种方法与制造天然石类似，但通常更为经济。

喷浆混凝土

喷浆混凝土是混凝土工程中一种特定的形式。在黏稠液体中特制的混凝土从一条封闭的路线运送到建设场地，并用气动枪喷射到备好的表面上，包括铸造形式、土壤或其他构件。这种冲击压力将会密封地面土壤。可能需要进行加固，这取决于如何使用水景设施以及基地的质量。

建议在改变地形的设施规划中使用这种方法，可为水景设施或需要经受高强度使用的装置设置一些与建筑结构的联系。如果建设场地很难进入的话可以利用管线运输。

塑料构件

塑料或者玻璃纤维合成树脂可用这种方法制造，并根据制造商的指示在场地中降低进入基坑。大型水景设施通常需要进行额外的回填，例如用少灰混凝土。塑料构件对于小型到中型的水景设施来说易于使用，有些可重复使用。但是很难隐掩玻璃纤维水池，使其从美学上与周围地面的结构整合（图 47）。

这种方法对于临时装置很常见，例如，装饰水池像地上游泳池一样，是一种用石头或完全可见的东西来隐掩的地下构造。

钢结构

钢结构用于喷泉装置、装饰性水池、克耐普式足浴、游泳池以及将长期大量使用的水景设施，如水坝、水槽或游玩区。钢的光学软度总会令人惊奇（图 48）。它具有高度耐用性，并且可以精确地安装和使用，若要根据特性和精确的要求建造一个倒映水池时，钢材是非常实用的。

对于防腐不锈钢的使用要格外注意，因为这种材料看起来是很单调的。这个问题可以通过多种表面处理方法来解决，包括喷砂、喷漆或在表面涂粉末涂料，但喷漆一段时间后会磨损并脱落，因而有碍观瞻。

镀锌和粗钢越来越多地被用来建造喷泉和堤岸，其他常用材料还有耐气候的钢材，具有适度腐蚀性的不锈钢合金，风化现象使表面“生锈”（图 49）。

木材

木材作为水景容器有着悠久的历史，特别是在森林地区。整个树干可以作为老式管道和水槽。长盒状的水渠，密集地放置在一起，是用于建造上射式的水磨、特色倒影池和巨大的喷泉装置（图 50）。

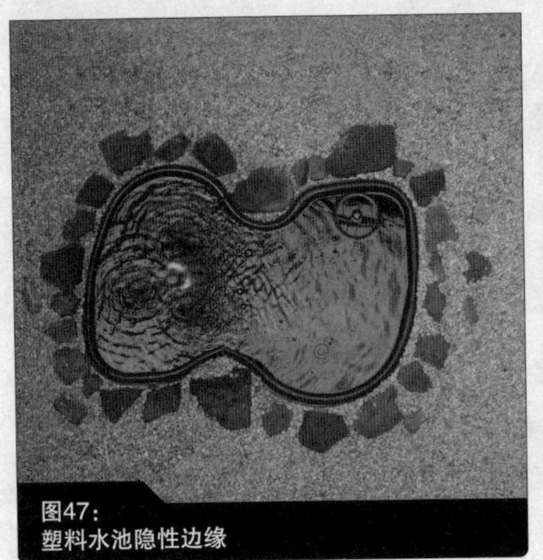

图47：
塑料水池隐性边缘

图48：
植物种植在薄壁钢槽中

图49：
喷泉厚厚的钢壁装有喷水头

图50：
最初的喷泉装置多为木质

提醒：

即使是耐气候的建筑钢材，如果长期处于湿润状态或是短期浸泡，也会被腐蚀。因此，推荐使用比结构所需还厚的墙，这种厚墙是为了避免浸湿过程中的损失。

图51：
有薄管的木地板细部

这些木材的使用可用于建筑感的形式，因此可以与混凝土的施工方法相协调。永久性建筑物应该使用高强度木材，它们也适合水利工程建设（橡树、落叶松和热带树林）（图 50，图 51）。

围合的水池

原生态或自然石材的围合水池在冬季气候温和的地区很受欢迎。围合的结构为形式的发展、混合材料的使用和设计结构良好的表面提供了很大的自由（图 52）。

然而，稳定的结构是必不可少的，用以防止由水体运动引起的裂纹，或可以使用低孔隙体积的抗霜砖，来防止因寒冷天气产生的裂缝。垫层的问题总会出现在许多关键环节之中，可以通过安装预制混凝土水池得到纠正。然而，这种方法将影响到水景外观的设计，适合小型人工池塘和渠道工程的使用。

天然石材

毫无疑问，天然石材是建设水池、石缝间的喷泉、水槽等最重要、最出色的材料。

混凝土基础上方结合一层自然石材面层，是天然石材利用的另一种常见方式。水渠由缘石、鹅卵石或粗糙的砾石构成。为了节省材料，降低成本，小型的喷泉石可以覆盖一层石制的面层，然而即使精准的节点处理和填充也无法消弭石材面层肌理二维，更确切地说是破碎的质感（图 53）。

为了避免面层石材的加工问题，经常利用砂岩来建造喷泉石。砂岩这种天然石头，通常是由石匠手工处理。天然石材被用于处理成更高层次的石材雕塑，这一点对于具有装饰功能的曲线型、碗状

图52：
喷泉池由天然的细条石建成，铺设砂岩，周围种有植物

型雕塑来说是尤为必要（图 54）。如果石雕技术高超并绝对精确，石头之间的接缝会极细微到能防水。灌浆铅是密封接头最持久的方法。

由单个石块组成断断续续的石景，与流动的水令人印象深刻（图 55）。整块的巨石设计，取决于采石场内的石层、石材初加工的程度和运输的限制。

天然石材可以适应最微小的变化，所以即使在水的体积很小时也可以精确地控制小瀑布与跌水的流量。表面处理也增强了整体的设计外观。研磨或抛光的表面突出了闪闪发光的水面，强调形式和优雅。斑驳或轻微喷砂工艺使表面看起来像天鹅绒般的亚光，最终会风化并随着时间增长变得越来越有趣。较粗糙的过程例如穿孔或粒化，不仅能给石头一个质朴的外观，也会根据水的深度减缓水流，让青苔在其表面生长。

提醒：

天然石材是自然侵蚀过程后的产物。现有的孔隙、细微裂缝、水穿透的石头、由霜冻引起的裂缝，以及由温度升降和时间引起的张力，都加速了天然石材的侵蚀过程。为了避免这一点，在水景设计中应采用制造商认证和保证过的具有良好的抗霜冻性和抗侵蚀性的自然石材。

图53：
表面贴了一层天然薄石板的方形喷泉

图54：
由几大块砂石模制拼成的喷泉

图55：
薄薄的水层流过巨大的天然石材喷泉

图56：
在临时花园内蜡质喷泉的一个少见实例

不规则的水池

对于较小的水景装置来说，可用的材料种类繁多，例如像池底垫层的设计和形式统一的喷泉和水槽，或者那些着重表现设计的概念的装饰性设计水景。区域环境的文脉、当地产出的原材料、传统工艺和方法，都适用于这种水景方式的表达。玻璃、陶器或蜡（图56），铝、铸铁、青铜、铅和其他金属和合金，以上这些材料都可在不规则的水池中进行应用。

P51

水岸设计

水岸设计是创建水景设施最重要的步骤之一。水岸类型的选择与开发方式应具有整体连贯性。它的形式——可以是利用自然植被和人工堤岸进行分隔的，也可以是用栈桥、入水踏步或亲水平台的参与式的——影响水景设施的使用和体验的方式。水岸设计体现了水景设施的整体特征是自然的还是人工的、是实体式的还是构架式的。

位置

水岸通常是一个水景设计中最敏感的区域，因为它是水体到周围环境的过渡。一方面要吸引游览者，创造尽可能多的亲水体验；另一方面，水岸区经常暴露在有水位波动、岩石水渗透或水波稳定的涨落之下，所有这些变化都需要由细部设计和精确的构造来实现。

水岸设计应保持连续，并且应该不低于设计的最高水位线，最高水位线是由水的注入和溢流决定的。水岸必须与池底的形式完全适合并相融合。水位的波动是由蒸发或使用引起的，在设计中也必须加以考虑。

毛细效应

大多数材料都有各自的毛细效应，这是由材料的内在毛孔结构所决定的，并呈现出由强到弱的过渡趋势。这种毛细作用使水越过可见水面渗入土壤，导致水池中的水流失。渗透进入土壤的水持续浸泡水岸，导致水岸的稳定性降低，这种降低依据水岸材料的毛细效应各有不同。

水位差

水位差通常在水岸设计中用以防止堤岸的破坏和水的流失，对于带有伸展到最高设计水位的水岸结构来说，水位差是一种安全的保障（图 57）。水位差高度的决定因素有：水景的规模、显露程度、使用方式及水体的冲击力。较小的功能设施就不太需要考虑这个问题，比如小型水钵或鸟类的浴池，流水溅在两侧是没有问题的。中型的公园池塘，建议水位差高度在 20 ~ 40 厘米。对自然水体来说，水位差可以延展到 1 米。

提醒：

在水岸设计中加入隔水层有助于阻断毛细作用，因此可以防止水的流失和水岸区的破坏。比较适宜的选择包括混凝土或在边缘放置低孔隙率的天然石材，同时松散地填充具有低细颗粒含量和大孔隙率的粗砾石或碎石砂。

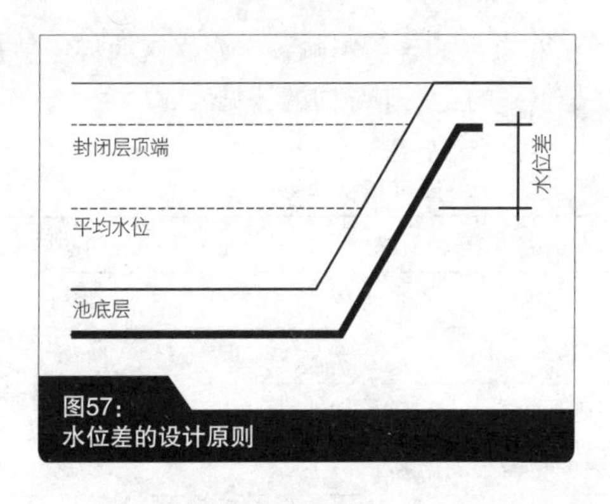

图57：
水位差的设计原则

水位差使水岸有更坚实的外观，并在水的体验和游览者之间创造了一个分隔。但这使水面在整体的比例关系中看起来过深过小，这种印象可通过三方面手段进行缓解：柔软的植被护岸、可见的池底构造、引入水中的踏步。

堤岸

水岸由池底的材料决定，并通过不同的形式、材料和构造工艺，形成一个连续的、形式上有结合力的、耐用的池底（图 58）。水岸很脆弱，因为水的运动会对其冲刷和侵蚀，高强度地使用而产生的升温或震动引起的破坏，会在设施周围产生裂缝，（例如栈桥）或在接缝处使材料发生改变。

形成水岸的一种简单的方式是缓缓将池底延伸到池壁的位置，最终到达水位差顶端。这种延展性材料，例如石油沥青砂胶或泥浆池底，可以保证坡度最大化。并且当需要考虑湿度时，这种最大化的堤岸形式能够保证最低水位的稳定，创造相对较宽并带有狭长浅水区的堤岸。运用混凝土墙或包裹的石头形成的水岸，允许有陡峭的斜坡、狭窄的堤岸和深水区。

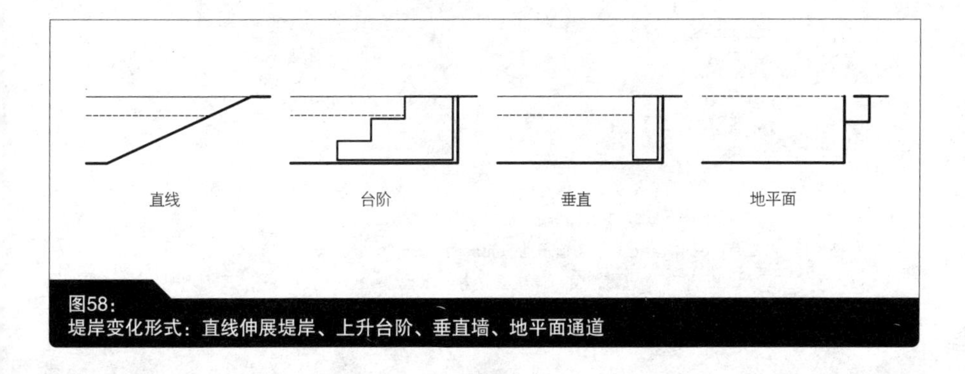

图58：
堤岸变化形式：直线伸展堤岸、上升台阶、垂直墙、地平面通道

材料的选择和水岸的走势影响一个水岸区的特征是开放的、自然导向的，还是整齐的、人工导向的（见表 3）。

堤岸可用材料一览表 **表 3**

开放的、自然导向的堤岸的推荐材料	整齐的、人工导向的堤岸的推荐材料
草地	墙
树篱、灌木	阶梯
砾石、砂	通道
包裹的石头 / 岩石	石笼
柳条	

风景优美的自然水岸

自然式堤岸由水深、边缘宽度、材料和植物起伏的相互作用，因此需要一个变化的多样的水岸。在选择池底的类型（参见“技术参数”一章中的“池底垫层和水池”）、水岸区的形式、水位的微观整合，或在为水岸植被创造合适的栖息地时，不规则的外观给予了高度的弹性（参见图 59、60 和“技术参数”一章中的“植物”）。

建筑化的人工水岸

如果希望效果显而易见，硬质的、人工的水岸，如水景墙、喷泉水池或下沉式的水池护岸，需要进行精确的规划和建设。这种方法需要具有低尺寸容差的材料（如天然石材或混凝土构件）、精确的给水、排水元素和技术支撑（如循环泵）（图 61、图 62）。

P54

进水和泄水

每个水景系统都需要进水，来维持水位的日常平衡，泄水一般是从池底自然流出，如果可能的话，会通过水装置排干。

提醒：

可见的表层水质受流入水体的调控。一种方法是低表面张力的给水管路，在水下深处和与上游连接的水塘上面，创造了清澈光滑的水面。水道冲刷着水体边缘，由中度表面张力给水管道，制造出带状浪花和环形浪花（出自同一个水源）。

图59：
水流量较低的草沟

图60：
浅齐的砾石滩散布着大量砾石，延伸到水中

图61：
硬质铺装，现浇混凝土质地的弯曲边缘

图62：
地平面高度的反射池细部，带有精细的钢磨边结构

进水和泄水是任何水景设计的开端和结束，要遵循整体设计的概念。水景装置的液压能力必须作为一个整体，与系统相对应，它可以不需要加压系统、独立操作，也可以使用水泵和高架罐的加压系统。这些系统可以隐瞒或显露出来，应保持到便于维修工作和保护的形式，以防过滤系统内的有害杂质。

隐蔽式进水和泄水

隐蔽式进出水系统由压力软管和吸水软管组成，这些都集成在一个水泵系统内，隐藏在水的表面下。另外一个水池位于进水口的前方，可以减缓水流的速度，以控制、减少或避免水面的漩涡（图63）。作为减压系统，使用水槽、水渠、进水管、进水和泄水的过程可以藏在水体外部。

可见式进水和泄水

由于可以将进出水系统可视化，带有设计感的进水和泄水系统是丰富多样的。例如，它们被设计为水面上一个突起的简单图案，或作为不断涌动的喷泉。溢流形的水坝、沿岸的水泵装置、技术静谧的输水系统，可以通过精心的设计结合在环境中。另一个出水口

图 63：
静止水面几乎看不出建于池底的进水口

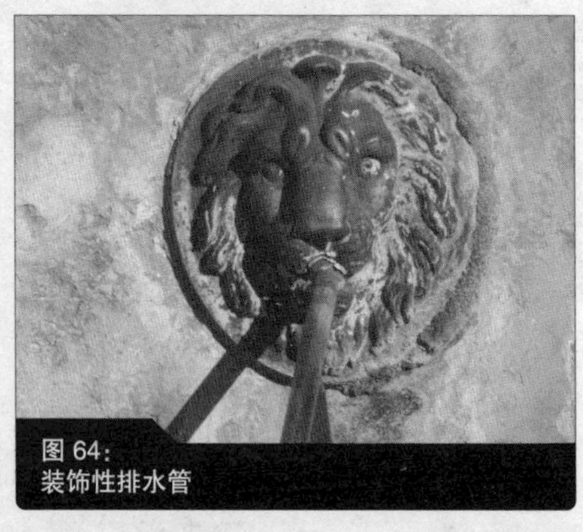

图 64：
装饰性排水管

图 65：
由陶土烧制的出水口

图 66：
高出池塘水面的叶子状保护格栅

被延伸的简单例子（图 64），这类设计可以由铸铁或青铜制成，或装饰性强，或简洁大方，或直或弯。水可以从构筑元素中涌出，例如天然石材、人工石、水坑、水槽、钢格板等；涌出过程中，水可以产生浪花或保持平静。最终，通过运用雕塑的形式表现出来，如：雕塑感很强的人工元素，寓言、童话中的人物，或圣人形象的喷泉等，这些所有的喷雾、喷泉、跌水通过集水排水沟收集起来，并加以设计来强化这种效果（图 65 ~图 67）。

泄水孔

为了保养工作和冬季排空水池，（技术参数这章——冬季防护）池底的泄水孔应被设计为一个简单的技术装置，隐蔽并易于操作。该装置也可以是一个泄水单元，安装在池底最深的位置，用滑动阀门就可使其开合，类似于船底排水泵，也可作为整个泵系统的一部分，或在较小的装置中，当排空水池时这一简单的溢水竖向管可以完全移动。

图67：
在反射池中的出水口

P57

水体流动

水体的流动影响着水体显性特点。流动水体的人工装置可以分为三类：流水、落水、喷水。

流水

流动的水体像小溪或沟渠一样，由高到低流下。它们也可以是简单的沟渠形式，一连串的水坑，或是自然与人工的结合，以线性或蜿蜒的形态出现。沟渠需要长期给水，如果中断了就会变干。明智的做法是巧妙地保留一些水，当整个系统停滞或持续干旱时，采用内置障碍和凹陷地形的方法来保持活力。

水的总量需要一个计算装置，应用出水口的直径、斜度、河道表面粗糙程度以及水的流速来计算。水随着均匀的坡度流下，坡面宽阔的部分会减缓流速，而狭窄的部分会增加流速（图 68）。水面上波浪和漩涡的外观受河道走向的影响，它转化的频率由水流动的速度、现存的谷底和裂缝，以及表面的结构来决定（图 69）。

落水

瀑布用于自由坠落的水体。对于这种方法，水在坠落前先被收集起来，减速并被保存，之后流过突起物，从边缘落到池底。

提示：

对于进水和泄水系统的连接，一种是用简单的技术实现水体的缓缓流动，设计时重点部分要随着自然水体流动而减压；另一种，活泼、旋转的水面需要足够的压力去推动水体。

提醒：

如果在平静的池水内安装一个高强度的瀑布，水表面的反射将延续（扩散）到水中。为使静水池内保持平静，紧连着冲击形成的瀑布可在池水中设置一块石头，这会中断水面上明显的运动并保持水面平静。

图68：
水渠设计中，水体流过雕塑般的表面

图69：
水坝营造出跌宕和变速的水流

悬挑的高度、峭壁的边缘、水流的速度决定各级水链的清晰度、持久性和平缓程度（图70）。越是要达到平缓、清澈、连贯的瀑布，就越是要精确计算出流动水体的常量；并校正悬挑和边缘的表面、结构和水平装置。天然的、未被开发的瀑布是不断变化、湍急、雾气缭绕的。它们在运行时具有不规则的边缘、变换的水流，冲刷石头以加强喧闹的气氛。由于预先确定落水实际的光学效果不是件容易的事，最好在建造阶段试运行，以改善和调节。

特别是那些计算精确的瀑布，表层水量明确，残余的水能够被灵敏的反应出来。即使是轻微的水杂质，例如树叶，也足以撕裂这层水膜，使落水变得不规则。好的场所，在上游悬崖边形成的屏障或过滤器可以防止这些问题，并使落水清澈无杂质。

喷水

喷泉由喷涌到不同高度和强度的喷嘴组成。他们可以单独或成组设置，射程和运转可以恒定不变也可以有节奏的变化。特殊组合、水压、不同形状的喷嘴与固定装置形成的喷泉各不相同（图71）。

单射程喷嘴（图72）可以营造出一个清澈、防风的效果；满水的喷嘴射程可高达14米，坡度达15°。多射程的喷嘴营造出聚集或分散的水体。当被安装在一个旋转的基地上时，反冲的水形成旋涡的形式。节水形式的空心喷头用于那些高达80米的喷泉。水膜喷嘴营造出封闭但随风摇曳的形式，就像圆顶状的水钟。扇形的喷嘴营造出不可思议的30°角，整幅密闭扇形水雾。手指状的喷嘴营造出垂直或倾斜的、异形的或是表面破碎化的水体。如果将空气加到

图70：
水花四溅的水层

喷头里，喷雾就与其他形式形成鲜明对比。气泡，并带有高强度防风的喷雾体，精心设计的切换机制和快速加压装置让水体呈现摇曳、舞动的形状（图 81）。

户外雾状喷泉装置是一种特殊形式的喷泉设施，它使水体应用高压装置可以很好地喷射出。优良的喷雾会降低空气温度，在夏季提供新鲜空气。它产生出不断变化、转瞬即逝的形状，在透明与不透明之间魔法般的变换与波动。装置优良的喷嘴对有杂质和含钙的水敏感，水分流失的多少根据风吹的情况而不同，因此喷泉需要连续地进水（图 73）。

提醒：

如果水的流速过低，由于水的黏附能力将阻碍清澈溪流的形成，甚至阻碍水体自由坠落的转化。水体在坠落的过程中“被卡住”，又游荡回来，或是落到偶然形成的支流中。因此，在石头边缘的下方设有回水槽，可以避免这一问题。

提示：

在设计喷泉时，考虑水池的合适尺寸是很关键的，这样风吹走的水就能被收集回来。根据风强度的不同，水体深度与水面宽度的推荐比值在 1 ∶ 2与1 ∶ 3之间。在狭窄的水池里，导致相邻设施的使用困难，这一点可用恰当组合自动风计量控制系统来纠正。

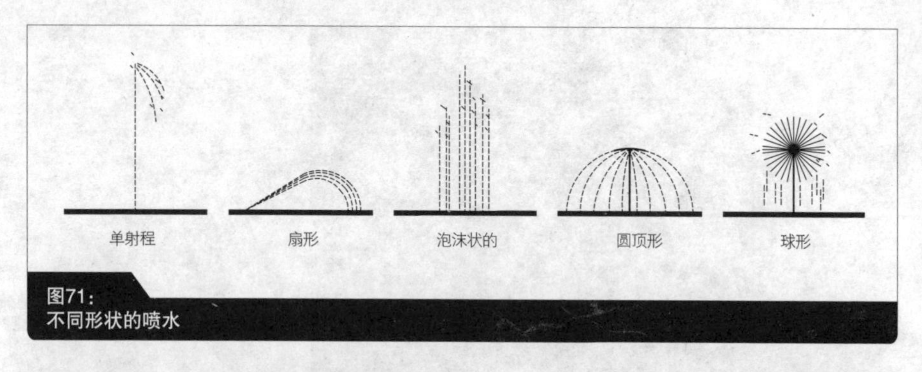

图71：
不同形状的喷水

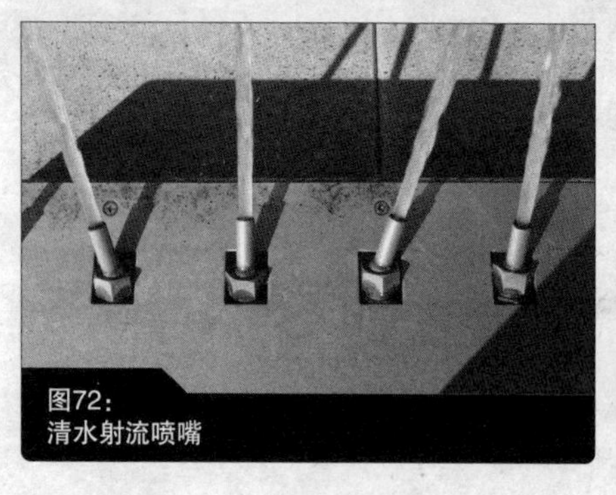
图72：
清水射流喷嘴

图73：
带有临时水雾的水上娱乐区

P61

水泵和技术

水的运动是水体在不同高度与压力下寻求平衡而导致的结果，它由水泵、管道系统、蓄水池，人工地创造出来。

水泵的类型

浸没式水泵被安置于水体内水池的中央，以节约空间，或是作为保护水体内部附近的分水岭，而被固定在一个平台上，这些装置可以浮在湖中央的水面上，并被用来设计大型喷泉装置。

独立式水泵可以被留在水体外面单独的水泵室中，用电缆与水库相连。独立式水泵的集合更为复杂，因此也更昂贵，但维护容易一些，使它相对来说比较合算。独立式水泵通常用在大型的、公共水景装置中（图 74）。

装置

水泵系统水泵应用防护格栅来加以保护，以防范事故，用过滤器、砂砾收集器避免杂质的损害。水泵由浮动开关或磁性开关、电脑程序自动操作。

装置的型号与尺寸，规划好的运转、水的体积决定了水泵的选择。对于较小的简易装置，广泛应用的预制方法一般是充足的。水泵严

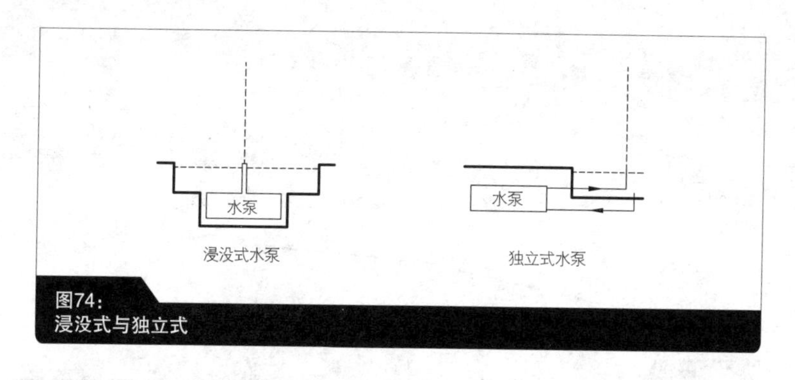

图74：
浸没式与独立式

重影响装置的运行成本，当考虑该设计时，从专业工程师和制造商那里找建议是明智的。

P63

照明

黑暗会削弱水对光强的吸引，瞬间把水变成一个黑暗的、无法穿过的表面。照明设计可以用特定的照明进行渲染，使水在夜间也充满生气。

外部光源

光从外部打到水面上，在黑暗的水面创造出一种舒适的映像。映射强度取决于光源的亮度、光与水面之间的距离、光谱以及水体表面的流动。颜色可以由光源添加，但它们的映像看起来很苍白并且饱和度较低。

水下光源

强光可以通过把光源置于水中来实现。其位置可以沿洼地周边，或在土壤中，或靠近喷泉。根据照明的亮度、到周围固定位置的距离、水的透明度及水池墙壁的反射可以照亮整个水体。由于水体流动而产生不同角度的折射，光产生不断变化的结构模式，并将光影的闪烁投射到周围环境。

提示：

在没有蓄水池的瀑布或流水装置中，池底的水泵最终会收集其余流动的水，所需的水体可被储存在可视的水池内，按等级流动，尺寸和周长要合适。隐蔽的、地下收集和补偿的水槽可减少形态效果，因此是设计时的首选。

示例：

以前的装置，例如德国卡塞尔公园内的水幕喷泉，大量使用了高架的水池，泵功率低但恒定，之后将水池排空，在那段时期是一项壮举。当时的技术既不能在任何其他时间操作，也不能预先确定的时间间隔。现代喷泉、水泵可设计为特殊的造型，灵活性强，可昼夜不停地控制和操作。

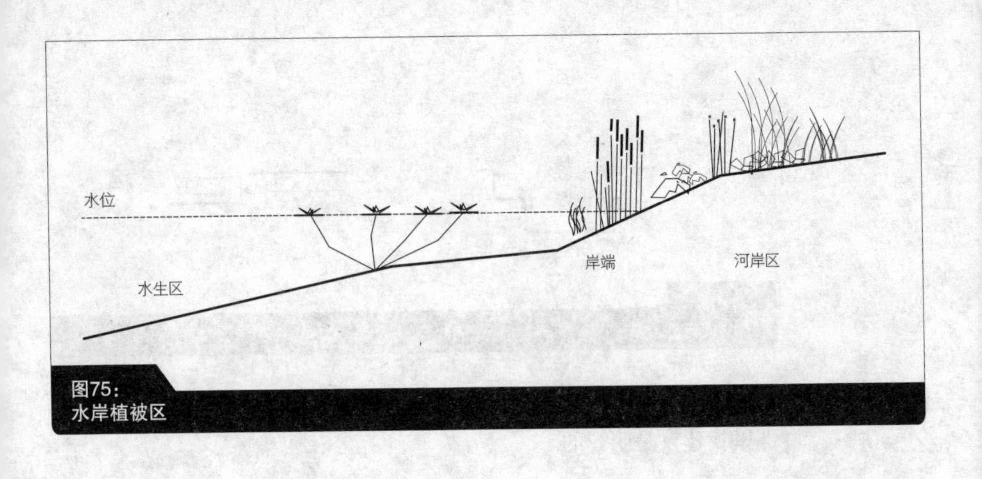

图75：
水岸植被区

只有特定的水下照明或冷色光照明才可用于水下照明设计中。在此过程中，外部投射灯的照射光通过玻璃纤维直射水中，可以通过使用彩色滤镜添加色彩。

P63

植物

人们头脑中关于水的图景几乎都是包含植物的：莫奈花园的睡莲、池边如画的柳枝、海滩中柔软的、沙沙作响的芦苇。即使是建筑化的水岸设计，人们也乐于通过轮廓缓和、形态舒展的植物来打断其严格的线性和精确的形式。

植物的高度、密度、颜色和叶片结构在一年的过程中不断变化。乔木可以在大空间中展示其令人印象深刻的魅力，如公园或风景优美的场地；在较小的空间中主要是利用灌木在细部强化景观特色。

选择位置

植物只会在适合其生境的位置茁壮成长，栖息地包含着植物种类的具体需求，如光、土壤和水等，例如柳树或赤杨比灌木更能在艰难的条件下生存，灌木对于环境中的变化和波动会有较敏感的反应。

植物的生境可根据其各自到水的理想距离划分为不同的区域（图75）。河岸区通常紧邻水的边缘，毛细效应使地面保持潮湿，也可能出现偶尔的淹水现象。对于花灌木和沼泽植物来说，这是一个完美的栖息地，它可以保持持续的湿润状态。挺水植物可以在深达60厘米的水中生长，这取决于他们的种类。对那些扎根于土壤中但有叶子浮在水面上的浮水植物和沉水植物来说，较深的水域是更加理想的生境。

图76：
柳树在每种文化中都是水体和河岸的标志

图77：
在一个规则几何形状的倒影池中下沉筐中种植的睡莲

植物对于水体运动变化是很敏感的，这种变化会限制植物的生长。如果水的流动过于强烈，叶片边缘会产生弯曲，水下植物会被连根拔起。从喷泉中溅出的水滴落在叶子表面，使日光产生一种放大镜的效应，会对植物造成“灼伤”。

生长力

给水生植物以适宜的营养和适度的水量，则其会以惊人的生命力茁壮生长，短时间内就可以改变水景的外观。水生植物过于繁茂会抑制其他种类的植物的生长。为保证水生植物的稳定性，所需的条件有：规模足以容纳种植区，植物基底缺乏营养，植被种植在限制其生长的筐中，水景设施定期维护（图 77）。

如果希望正确地选择与组合植物，形成一个有吸引力的、稳定的、可见的植物群落结构，则需要适合的植物种类和密度、适宜的养护和管理。

› ✎ › 🖇

✎

提醒：

自然存在的植物通过生物法则净化水体，这也很大程度上影响了我们的视觉体验，以至于我们常常在靠近水的地方看到的那些植物，如芦苇枝，甚至可以用来象征水。这些都可以技巧性地用来作为设计的元素（图 76）。

🖇

提示：

附录表 1–3 中提供了水生植物种类的基本指南。更多信息和设计方法可以参看本套丛书中的雷吉娜·埃伦·韦尔勒（Regine Ellen Wöhrle）与汉斯－约尔格·韦尔勒（Hans-Jörg Wöhrle）编著、齐勇新翻译的《植物设计》一书（中国建筑工业出版社，北京，2012 年）（征订号：21246），也可以参看 Richard Hansen 和 Friedrich Stahl 合著的《Perennials and Their Garden Habitats》一书（剑桥大学，1993 年）。

图78：
厚厚的海藻覆盖整个水面

P65

水质

尽管越来越多具有审美要求的案例集中于污染、变化和改造，但还有一些其他的案例可以展现出我们所希望的清洁、新鲜、透澈的理想水景。浑浊的水、藻类，甚至是苦咸的、恶臭的水体不仅表现出生物平衡性的缺失，也是设计上的失败（图 78）。

生物平衡

生物平衡是水体得以清洁和透澈的前提条件。只有基底的生产者和消费者——即植物、动物和微生物——彼此成比例时才可以实现。如果水面被阴影适度的覆盖；或有能带来氧气、水的流动；或水体更大、更深、更冷；并且没有被高强度地开发利用，具有较少数量的动物，这种理想的水体也可以实现。

在封闭系统中，生物平衡的实现需要通过天然的和人工的方式不断地改变与清洁水体，如利用电子设备或化学过程。

天然净化

大型的自然水景设施，如水面与周长具有均衡比例并有自净能力的游泳池，几乎完全使用天然的净化方法。可以在一个人工湿地中净化水，人工湿地内有芦苇、有时是香蒲、泥浆、或莎草。植物的根茎对于微生物来说是一个完美的栖息地，微生物能够降解营养物质和有害物质，给水提供氧气并有助于水的净化。

支撑天然净化能力的混合动力系统通过机械使水重新循环，并建议沿着水岸边缘区使用滤沙器，如公众泳池。

人工净化

人工净化模拟自然过程，但主要集中在小空间。在第一机械液压阶段，大块的杂物用纱布从水中移除，细微的浮动沉积物用撇油器、筛子和水晶石英砂过滤。在接下来的化学阶段，水的 pH 值通过添加酸进行中和或利用 pH 调节剂过滤。最后，细菌被具有杀菌功效的产品消除，例如氯（但在户外使用时是有争议的），或通过加入过氧化氢或氧化银来消除。

对于没有植被的人造水景设施，因为不具备天然净化能力，通常需要完备的技术设备，例如大量使用的水上游乐区，需要卫生的、高品质的水处理装置。

P67

安全

水元素一直兼具吸引力和威胁性。因此用水作为设计元素时对安全性的要求非常高，以防止发生意外的风险。可能出现的不同危险包括深水、在无人监督的情况下进入水中、水下技术设备的威胁，如水泵或进水管。

预防措施

水景设施建成后所必须融入的安全措施几乎总是对设计不利的，因此在设计阶段解决这些问题，并把对安全方面的考虑作为整体概念的一部分，就更有意义。考虑要素包括：

—控制亲水通道：封闭的与可接近的河岸应该清晰地表示，以便游客可以容易地区分两者。可接近的河岸应通过坡道、台阶、汀步或扶手等形式以使人与水可接近。

—合适的水深：在儿童玩耍的区域，安全的水深是非常必要的。如果是深水区，可以安装一个水下格栅以形成一个安全的、相对较浅的深度（图 79）。

—良好的逃生路线：如果有意外情况发生，应该在不需要其他人的帮助下就可以容易地远离水。这可以通过一个相对平缓的 1/3 或更小斜率的缓坡和踏步来实现。

—设备安全：技术设备如进水管或潜水泵应上覆格栅，以防范事故的发生。

P68

冬季防护

根据寒冷地区的季节性问题，即水能结冰并能开发出特定的冰雪审美效果的时间段，最多可以涵盖半年。由于水的异常，即冻结

图79：
安装在较深的水体表面的安全格栅

时其体积将扩大10%，因此低温将对水景设施的技术设备构成威胁。为了保护它，大多数水景设施在冬季开始时就被清空并关闭。与永久设施相比，它们看起来被废弃，暗淡无光，而且丑陋。好的设计总会从设计和技术角度考虑到设施全年的外观及其封闭阶段（如冬季），进行合适的设计来应对。

技术延迟

静态水的结冰速度比动态水要快。对水和蓄水洼地进行加热，或加入化学品，可以防止或至少延迟水的冻结，但建设维护费用较高使这种解决方案缺乏吸引力。

清空的设施

有霜冻风险的设备，在霜期开始之前，通常会被关闭并将设备完全清空，在此期间，应保持一个出口处于开放状态，以排除沉淀、渗流、凝结累积、会冻结的水。

简单的构造可在冬季关闭且无须保护，重要的结构如带有细部雕塑的喷泉和装饰性的元素，主要是通过木质外壳来保护。

填满的设施

带有池底的水景设施通常对于干燥的状态比较敏感，如黏土或淤泥，因此在冬季它们仍然需要注满水。在其周边还需要有足够的空间允许冰的膨胀，以避免损坏这些设施，现存封闭的并充满水的管道必须加以保护以抵抗冻胀。冻土的深度取决于当地的气候，平均冻土深度范围为0.8 ~ 1.2米。

水生植物已根据区域环境正确选择，由于它能够适应气候，故无须额外的冬季保护。为了确保水生动物的存活，冰层下面必须有足量的防冻保护水层，以提供尽可能多的栖息地。

喷泉的清理

冬季结束紧接着就是紧张的清洁期，移除保护套管，清除树叶、泥土和垃圾，重植繁茂的植物。在重开设备之前，对设备进行必要的检修。

P69

成本效益

景观设计的三种设计元素——植被、地形和水，其中水对景观维护成本有长期的影响，相比较而言，初始费用显得不那么重要了。中等规模的人工水景设施的前期建造成本和后期的维护成本基本一致。因为它们尺寸相近所以生产更加精美又生动的喷泉比较合理。

由于不间断的维护，使水景设施在中期的时候成本昂贵。经济因素可能会使好的设计的难以实现，或者可能为达到技术上的完善，在建设时由于缺乏资金而被关停。

因此，在规划和概念阶段，就成本而言，应尽力建立一个持久的、低成本维护的可持续设计。利用现有元素的可能性进行开发、降低必要的技术标准，使水景设施达成相对完善的技术和结构。

优化概念

从概念上讲，依赖于基地的解决方式，是选择较好的位置，最大化地利用协同效应和成本。被掩埋的布鲁克斯河可以被重新挖掘并开放，以一种全新的方式利用现有的蓄水洼地，使雨水通过开放的渠道自由流动，进入水箱和池塘（图 80）。

相应的标准

非精确化的设计以及略有降低标准的水景设施，更容易实现和维护。这种设计需要考虑到成本相对低廉的自然过程，例如近天然的游泳池是依靠自净能力，而不是像传统的游泳池那样需要复杂而昂贵的技术设备。

示例：

2002 年，德国 Grossenhain 一个室外游泳池被替换为天然游泳池。2000 平方米的更新表面邻接 3000 平方米的泳池，每天有高达 1800 名游客使用。相比于传统的游泳池，该天然游泳池的运营成本降低了 40%。

图80：
收集雨水的水池，植被和木栈道形成了公共广场的焦点

图81：
卢浮宫前的水景

优化结构

为避免后续维修的需要，建议选择一个持久的解决方案。应利用任何可能的简化和减少维修及保养成本的方案。例如减少使用维护成本过高的设备，或使用维修和拆装方便的构件，或尽量使用同一运营商的设备，以保持其操作系统的完整性和连贯性。

人工水景设施的注意事项有：

—蓄水洼地应尽可能浅，以使填充和换水的成本最小化；
—水泵应按照正确的要求单独安装，以优化能源和水的使用；
—喷泉不应设在有风的通道，从而避免过多的水分流失；
—全面的冬季保护措施将防范技术设备的损坏。

自然水景设施的注意事项有：

—水体应尽可能大，以避免水温变暖的速度过快；
—植被应种植在生物篮中以控制它们的生长；
—不应引入鱼类，因为它们所带来的营养物质会促进藻类的生长。

P72

结论

在设计室外空间时，水是必不可少的、几乎是无法避免的元素。其象征性、活力、独特的情感品质以及其内在的多样性为开发适宜的、独特的、带有视觉诗意的整体概念提供了实现的可能。水的设计吸引着人们的注意力——如何自然地运作，人们如何感知其运作，以及它如何带动观察者的情绪。人与水接触所触发的与整体意境相关的深思，对场地独特性的解读、材料的准确性、工艺的品质，以上这些都可以作为一个设计是否成功的依据。

水是自由的。在物理规范的界限内，有一种几乎无限量的空间创造力。水能引发好奇心和体验的愿望。它能提供长期持久的吸引力，这始终令人惊异。本文旨在简要的介绍——激励大家去观察，号召大家来体验，并鼓励探索新的设计途径。

P73 # 附录

P73 ## 水生植物

下表提供了一个常用水生植物的概述，并且可以作为植物种植的入门指导。

湿地灌木选择　　表 1

品种	中文名称	光照要求	植株高度	花期（月份）	花色
Ajuga reptans	紫唇花	半荫	10~20cm	4~5	蓝色
Caltha palustris	驴蹄草	全光到半荫	30cm	3~4	金黄色
Cardamine pratensis	草甸碎米荠	半荫到全荫	30cm	4~6	粉—淡紫色
Carex elata "Bowles Golden"	黄金立苔	全光到半荫	60cm	6	棕色
Carex grayi	筛纲藻苔草	全光到荫蔽	30~60cm	7~8	青褐色
Darmera peltata	雨伞草	全光到半荫	80cm	4~5	粉色
Eupatorium species	紫苞佩兰	全光到荫蔽	80~200cm	7~8	深粉色
Filipendula ulmaria	旋果蚊子草	全光到半荫	120cm	6~8	米色
Frittilaria meleagris	花格贝母	半荫	20~30cm	4~5	蓝紫色
Geum rivale	紫萼路边青	全光到半荫	30cm	4~5	金褐色
Gunnera tinctoria	洋二仙草	全光	150cm	9	红色
Hemerocallis wild species	萱草	全光和半荫	40~90cm	5~8	黄色 / 橙色
Iris sibirica	西伯利亚鸢尾	全光到半荫	40~90cm	5~6	蓝紫色
Leucojum vernum	雪片莲	半荫到全光	20~40cm	2~4	白
Ligularia species and hybrids	新疆千里光	全光到半荫	80~200cm	6~9	黄色 / 橙色
Lysimachia clethroides	狼尾花	全光到半荫	70~100cm	7~8	白
Lythrum salicaria	千屈菜	全光到半荫	80~140cm	7~9	蓝紫色
Myosotis palustris	沼泽勿忘我	全光到半荫	30cm	5~9	蓝色
Primula species and hybrids	报春花	半荫到荫蔽	40~60cm	2~8	白色 / 黄色 / 粉红色 / 橙色 / 淡紫—蓝紫色
Tradescantia- Andersoniana hybrids	无毛紫露草	全光	40~60cm	6~8	蓝色 / 蓝紫色 / 白色 / 胭脂红色

续表

品种	中文名称	光照要求	植株高度	花期（月份）	花色
Trollius euro-paeus/Trollius hybrids	金莲花	全光到半荫	40~70cm	4~6	黄色 / 橙色
Veronica longifolia	兔儿尾草	全光	50~120cm	7~8	蓝色

水岸灌木选择　　表 2

品种	中文名称	光照要求	水深	植株高度	花期（月份）	花色
Acorus calamus	水菖蒲	全光到半荫	0~30cm	最高 120cm	5~6	黄色
Alisma plantagoaquatica	水芭蕉	全光到半荫	5~30cm	40~80cm	6~8	白色
Butomus umbellatus	开花灯芯草	全光到半荫	10~40cm	100cm	6~8	粉红色
Calla palustris	水芋	全光到半荫	0~15cm	最高 40cm	5~7	黄 / 白色
Caltha palustris	驴蹄草	全光到荫蔽	0~30cm	最高 30cm	4~6	黄色
Carex pseudocyperus	莎草	全光	0~20cm	80cm	6~7	—
Equisetum fluviatale	沼泽马尾草	全光到半荫	0~5cm	20~150cm	—	—
Hippuris vulgaris	马尾草	全光到半荫	10~40cm	40cm	5~8	绿色
Iris pseudocorus	黄菖蒲	全光到荫蔽	0~30cm	60~80cm	5~8	黄色
Phragmites australis "Variegatus"	芦苇（带有黄色条纹变化的）	全光	0~20cm	120~150cm	7~9	呈褐色 / 红色
Pontederia cordata	梭鱼草	全荫耐半荫	0~30cm	50~60cm	6~8	蓝色
Sagittaria sagittifolia	茨菰	全光	10~40cm	30~60cm	6~8	白色
Scirpus lacustris	灯芯草	全光到半荫	10~60cm	最高 120cm	7~8	棕色
Sparganium erectum	刺芒芦苇	全光到半荫	0~30cm	100cm	7~8	绿色
Typha angustifolia	窄叶香蒲	全光	0~50cm	150~200cm	7~8	微红的棕色花序
Typha minima	小香蒲	全光	0~40cm	50~60cm	6~7	棕色的钟装花序
Veronica beccabunga	婆婆纳	全光到半荫	0~20cm	40~60cm	5~9	蓝色

水生植物选择 表3

品种	中文名称	光照要求	植株高度	花期（月份）	花色
Callitriche palustris	沼生水马齿	全光到半荫	10~30cm	—	—
Ceratophyllum demersum	金鱼藻	全光到半荫	30~100cm	夏季	不可见
Elodea canadensis	伊乐藻	全光	20~100cm	—	—
Hydrocharis morsus-ranae	蟾蜍水鳖	全光到半荫	> 20cm	6~8	白
Myriophyllum verticillatum	狐尾藻	全光到半荫	30cm	6~8	粉
Lemna minor	浮萍	全光	> 20cm	—	—
Lemna trisulca	星芒浮萍	全光	> 20cm	—	—
Nuphar lutea	欧亚萍蓬草	全光到半荫	40~100cm	6~8	黄
Nymphaea alba	白睡莲	全光	60~100cm	5~8	白
Nymphaea-Hybriden, e.g.“Anna Epple”, “Charles de Meurville”, “Marliacea, Albida”, “Maurica Laydeker”	睡莲	全光	40~60cm	6~9	粉 / 紫红 / 纯白色 / 紫
Nymphaea odorata “Rosennymphe”	香睡莲	全光	40~60cm	6~9	粉
Nymphaea tuberosa “Pöstlingberg”	块茎睡莲	全光	60~80cm	6~9	白
Nymphoides peltata	荇菜	全光到半荫	40~50cm	7~8	黄
Potamogeton natans	浮叶眼子菜	全光到半荫	40~100cm	6~9	白
Ranunculus aquatilis	水毛茛	全光	30cm	6~8	白
Stratiotes aloides	水剑叶	全光到半荫	30~100cm	6~8	白
Trapa natans	菱角	全光	40~120cm	6~7	浅蓝
Utricularia vulgaris	狸藻	全光到半荫	20~40cm	7~8	黄

参考文献

Alejandro Bahamón: *Water Features*, Loft Publications, Barcelona 2006

David Bennett: *Concrete*, Birkhäuser Verlag, Basel 2001

Bert Bielefeld, Sebastian El khouli: *Basics Design Ideas*, Birkhäuser Verlag, Basel 2007

Ulrike Brandi, Christoph Geissmar-Brandi: *Lightbook, The Practice of Lighting Design*, Birkhäuser Verlag, Basel/Boston 2001

Francis Ching: *Architecture: Form, Space and Order*, Van Nostrand Reinhold, London/New York 1979

Christian Cajus Lorenz Hirschfeld: *Theory of Garden Art*, University of Pennsylvania Press, Philadelphia, PA 2001

Paul Cooper: *The New Tech Garden*, Octopus, London 2001

Herbert Dreiseitl, Dieter Grau, Karl Ludwig: *New Waterscapes*, Birkhäuser Verlag, Basel 2005

Edition Topos: *Water, Designing with Water, From Promenades to Water Features*, Birkhäuser Verlag, Basel 2003

Renata Giovanardi: *Carlo Scarpa e l'aqua*, Cicero editore, Venice 2006

Richard Hansen, Friedrich Stahl: *Perennials and Their Garden Habitats*, Cambridge University Press, Cambridge 1993

Teiji Itoh: *The Japanese Gardens*, Yale University Press, New Haven/London 1972

Hans Loidl, Stefan Bernhard: *Opening Spaces*, Birkhäuser Verlag, Basel 2003

Gilly Love: *Water in the Garden*, Aquamarine, London 2001

Ernst Neufert: *Architects' Data*, Blackwell Science Publishers, Malden, MA 2000

OASE Fountain Technologie 2007/08, http://www.oase-livingwater.com/

George Plumptre: *The Water Garden*, Thames & Hudson, London 2003

TOPOS 59 – Water, Design and Management, Callwey Verlag, Munich 2007

Udo Weilacher: *Syntax of Landscape*, Birkhäuser Verlag, Basel 2007

Stephen Woodhams: *Portfolio of Contemporary Gardens*, Quadrille, London 1999

P77

图片鸣谢

All drawings plus the illustration on page 8 (Latona Fountain in Versailles, LeNotre), 2, 3 (Bundesplatz in Bern, Staufenegger + Stutz), 6, 7, 8 (City park in Weingarten, lohrer.hochrein with Bürhaus), 10 (Landscape Park in Basedow, Peter Joseph Lenné), 11 (Park Vaux-le-Vicomte in Meaux, Le Notre), 12 (Fountain in Winterthur, Schneider-Hoppe/Judd), 13, 14 (Japanese Garden in Berlin, Marzahn, Shunmyo Masuno), 15 (Zeche Zollverein in Essen, Planergruppe Oberhausen), 16 (Villa Lante Garden in Viterbo, da Vignola), 18 (Spa Gardens in Bad Oeynhausen, agence ter), 19 (Bertel Thorvaldsens Plads in Copenhagen, Larsen/Capetillo), 20 (Fountain in Park Schloss Schlitz, Schott), 21 (Duisburg Nord landscape park, Latz and Partner), 22 (Spa Gardens in Bad Oeynhausen, agence ter), 23 (Cloister Volkenroda, gmp), 24 (Oberbaum City in Berlin, Lange), 25 ("Fuente" drinking fountain, Battle + Roig), 26, 27 (Allerpark in Wolfsburg, Kiefer), 28 (Botanical garden in Bordeaux, Mosbach), 29, 30 (Outdoor swimming pool in Biberstein, Schweingruber Zulauf), 31 (City park in Burghausen, Rehwald), 32 (City garden in Weingarten, lohrer.hochrein mit Bürhaus), 33, 34 (Paley Park in New York, Zion & Breen), 35 (Oriental Garden in Berlin, Louafi), 36, 37, 38 rechts (Mosaics in Saint-Paul de Vence, Braque), 39 (Villa d'Este in Tivoli, Ligori and Galvani), 43, 44, 45 (Cemetery in Küttigen, Schweingruber Zulauf), 46 (Promenade in Kreuzlingen, Bürgi), 47, 48 (Fountain in Dyck, rmp), 49 (Fountain in Bregenz, Rotzler Krebs Partner), 50 (Fountain in Valdendas), 51, 52 (Garden in Hestercombe, Jekyll), 54, 55 (Sundspromenaden in Malmö, Andersen), 56 (Garten der Gewalt (Garden of Violence) in Murten, Vogt), 59, 60 (Brückenpark in Müngsten, Atelier Loidl), 61 (Ankarparken in Malmö, Andersson), 62, 64, 65 (Fountain in Saint-Paul-de-Vence, Miro), 66 (City garden in Erfurt, Kinast, Vogt, Partner), 67, 68 (Villa Lante Garden in Viterbo, da Vignola), 69, 70, 72, 73 (Parc de la Villette in Paris, Tschumi), 76 (Chinese Garden in Berlin, Pekinger Institut für klassische Gartenarchitektur), 77 (Open spaces on Max-Bill-Platz in Zurich, asp), 78, 79, 80 (Inre Strak in Malmö, 13.3 landskapsarkitekter), 81 (Louvre in Paris, Pei):
lohrer.hochrein landschaftsarchitekten

Figure 40 (Elbauenpark Magdeburg, Ernst. Heckel. Lohrer with Schwarz und Manke), 53, 63 (Rubenowplatz in Greifswald, lohrer.hochrein):
Hans Wulf Kunze

P78　　作者简介

阿克塞尔·洛雷尔（Axel Lohrer），Dipl.–Ing.（FH），执业于德国慕尼黑和马格德堡的 Lohere. Hochrein 景观建筑事务所。